RNA Isolation and Analysis

P. Jones
Department of Biology, University of Essex, Colchester, UK

J. Qiu
CRC Beatson Laboratories, Glasgow, UK

D. Rickwood
Department of Biology, University of Essex, Colchester, UK

©BIOS Scientific Publishers Limited, 1994
First published 1994

All rights reserved. No part of this book may be reproduced or transmitted, in any form or by any means, without permission.

A CIP catalogue record for this book is available from the British Library.

ISBN 1 872748 37 6

BIOS Scientific Publishers Ltd
St Thomas House, Becket Street, Oxford OX1 1SJ, UK
Tel. +44 (0)1865 726286. Fax +44 (0)1865 246823

DISTRIBUTORS

Australia and New Zealand
DA Information Services
648 Whitehorse Road, Mitcham
Victoria 3132

India
Viva Books Private Limited
4346/4C Ansari Road
New Delhi 110002

Singapore and South East Asia
Toppan Company (S) PTE Ltd
38 Liu Fang Road, Jurong
Singapore 2262

USA and Canada
Books International Inc.
PO Box 605, Herndon, VA 22070

Typeset by Marksbury Typesetting Ltd, Midsomer Norton, Bath, UK.
Printed and bound in Great Britain by Biddles Ltd, Guildford and King's Lynn

LIVERPOOL
JOHN MOORES UNIVERSITY
AVRIL ROBARTS LRC
TITHEBARN STREET
LIVERPOOL L2 2ER
TEL. 0151 231 4022

RNA Isolation and Analysis

Books are to be returned on or before
the last date below.

7-DAY LOAN

23 SEP 2005

LIBREX

LIVERPOOL JMU LIBRARY

3 1111 00782 7932

Contents

Abbreviations

A	adenine
AMPPD	3-(4-methoxyspiro[1,2-dioxetane-3,2′-tricyclo [3.3.1.1$^{3.7}$]decan]4-yl) phenyl phosphate
APS	ammonium persulfate
ATA	aurintricarboxylic acid
BAC	*N,N′*-bisacrylylcystamine
BCIP	5-bromo-4-chloro indoxyl phosphate
BSA	bovine serum albumin
C	cytosine
CA	cellulose acetate
CAIP	calf alkaline intestinal phosphatase
CCD	charge-coupled camera
cDNA	complementary DNA
CDTA	cyclohexanediamine tetracetate
CSPD	5-chloro derivative of AMPPD
DAB	diaminobenzidine
DADT	diallyltartardiamide
dATP	deoxyadenosine triphosphate
dCTP	deoxycytidine triphosphate
ddNTP	dideoxynucleoside triphosphate
DEAE	diethylaminoethyl
DEPC	diethylpyrocarbonate
dGTP	deoxyguanosine triphosphate
DMSO	dimethyl sulfoxide
DNase	deoxyribonuclease
dNTP	deoxynucleoside triphosphate
DTE	dithioerythritol
dTTP	deoxythymidine triphosphate
DTT	dithiothreitol
dUTP	deoxyuridine triphosphate
ECL	enhanced chemiluminescence
EDTA	ethylenediaminetetraacetic acid
EtBr	ethidium bromide
G	guanine
HEPES	*N*-(2-hydroxyethyl)piperazine-*N′*-(2-ethanesulfonic acid)

hnRNA	heterogeneous nuclear RNA
hnRNP	heterogeneous nuclear ribonucleoprotein
HPLC	high-pressure liquid chromatography
HRP	horse radish peroxidase
HTAB	hexadecyl-trimethyl ammonium bromide
IMS	industrial methylated spirits
LGT	low gelling temperature (agarose)
MES	2-(*N*-morpholino)ethanesulfonic acid
MICAD	microchannel array detectors
MOPS	3-(*N*-morpholino)propanesulfonic acid
mRNA	messenger RNA
mRNP	messenger ribonucleoprotein
NMR	nuclear magnetic resonance
NTA	nitrilotriacetic acid
OD	optical density
PAGE	polyacrylamide gel electrophoresis
PCR	polymerase chain reaction
PEG	polyethylene glycol
PEI	polyethyleneimine
PIPES	piperazine-*N*,*N*′-bis(ethanesulfonic acid)
PMSF	phenylmethylsulfonyl fluoride
PPO	2,5-diphenyloxazole
PVS	polyvinyl sulfate
RNase	ribonuclease
RNP	ribonucleoprotein
rRNA	ribosomal RNA
RT	reverse transcriptase
S	Svedburg unit for sedimentation ($1S = 10^{-13}$ sec)
SDS	sodium dodecyl sulfate
SLS	sodium lauryl sulfate
snRNA	small nuclear RNA
snRNP	small nuclear ribonucleoprotein (‘snurp’)
T	thymine
T_m	melting temperature
TBE	Tris-borate-EDTA
TCA	trichloroacetic acid
TEMED	*N*,*N*,*N*′,*N*′-tetramethylethylenediamine
6-TG	6-thioguanosine
TLC	thin-layer chromatography
TMG	2,2,7-trimethylguanosine
tRNA	transfer RNA
4-TU	4-thiouridine
U	uracil
UV	ultraviolet radiation

Preface

The development of efficient methods for the isolation and characterization of RNA has been the key to the elucidation of the complex processes involved in the regulation of gene expression. The techniques used have varied over the years and most recently approaches using the polymerase chain reaction (PCR) methodologies have made a significant impact on work in this area.

This book begins with a broad-ranging overview of the structure and function of RNA and the subsequent chapters collate the diverse range of methodologies used in this area of research, from the isolation of RNA to determination of the sequences of individual molecules. All molecular methods in everyday use in molecular biology laboratories throughout the world are described. The emphasis is on techniques used for mRNA since this reflects the main interests of researchers, but many of the techniques used for mRNA can also be used for other types of RNA. Space limitations have prevented us from considering the increasingly powerful physical techniques such as X-ray crystallography now being used to analyze RNA molecules and we hope that readers will accept this omission. While intended primarily for novices in this area of research, this book should also prove useful for experienced researchers wishing to update their knowledge.

Paul Jones
Jane Qiu
David Rickwood
1994

1 RNA structure and function

1.1 Introduction

The roles of RNA within any cell are extremely diverse, with functions ranging from transfer of genetic information from the DNA to the protein synthetic machinery, to the enzymatic activities of ribozymes. The ability of the RNA to carry out this diverse range of functions is dependent on their sequences and their molecular modifications. The simplistic view that the sequences of RNA merely reflect the sequences of the DNA has long been discarded following the discoveries of the various post-transcriptional modifications that RNA can undergo. This chapter gives an overview of RNA as a macromolecule to provide a backdrop to the subsequent chapters on the techniques used in this area of research. For further, more detailed information, the reader should consult the further reading list given at the end of the chapter.

1.2 Chemical components

Ribonucleic acid (RNA) is a long and, usually, unbranched polymer made up of a linear array of monomers called ribonucleotides. Nucleotides are associated with two types of bases, pyrimidines and purines. Cytosine (C) and uracil (U) are called pyrimidines and they are simple derivatives of a six-membered pyrimidine ring; guanine (G) and adenine (A) are purine compounds, with a second five-membered ring fused to the six-membered ring (*Figure 1.1*). In addition to these four main bases of RNA, other minor bases are present (*Figure 1.2*); these are derived from the four main bases and are present in very precise positions in RNA, particularly in tRNA and rRNA where the modifications play an important role in the function of the molecules. The bases are linked to the sugar ribose and this in turn is linked to a phosphate group to give a nucleotide. The internucleotide bond in RNA is a 3′–5′ phosphodiester linkage. However, in some cases an unusual nucleotide linkage can be shown to occur, for example, in splicing reactions when branching occurs.

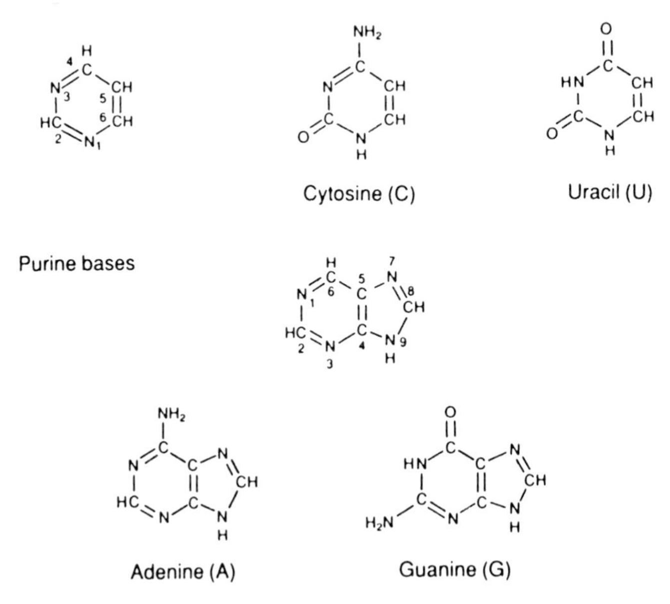

Figure 1.1: Bases of RNA. Reproduced from Chambers and Rickwood (eds) (1993) *Biochemistry Labfax*, p. 248, BIOS Scientific Publishers Ltd.

1.3 Secondary and tertiary structure of RNA

While RNA molecules do not possess the regular interstrand hydrogen-bonded structure characteristic of DNA, they have the capacity to form double-helical regions and actually show greater structural versatility than DNA in the variety of species, the diversity of conformations, and chemical reactivity. Indeed, the many biological functions of RNA, particularly tRNA, are based on very specific three-dimensional RNA structures.

Antiparallel double helixes can be naturally formed between two separate RNA chains but more usually they occur between two segments of the same chain folded back on itself. These short double-helical regions are connected by single-stranded stretches, adopting a globular shape. The secondary structure is similar to the A form of DNA with the bases tilted, since the 2′-OH hinders the formation of the B type helix. The helical regions formed in this manner are seldom regular. When non-complementary segments of the chain are brought into proximity, non-paired residues are looped out of the structure. In most RNA species, the secondary structure is dominated by such stem-and-loop hairpins. The helical hairpin regions can form because of their antiparallel orientation of some complementary sequences found in different parts of the RNA chains. However, structures of this type

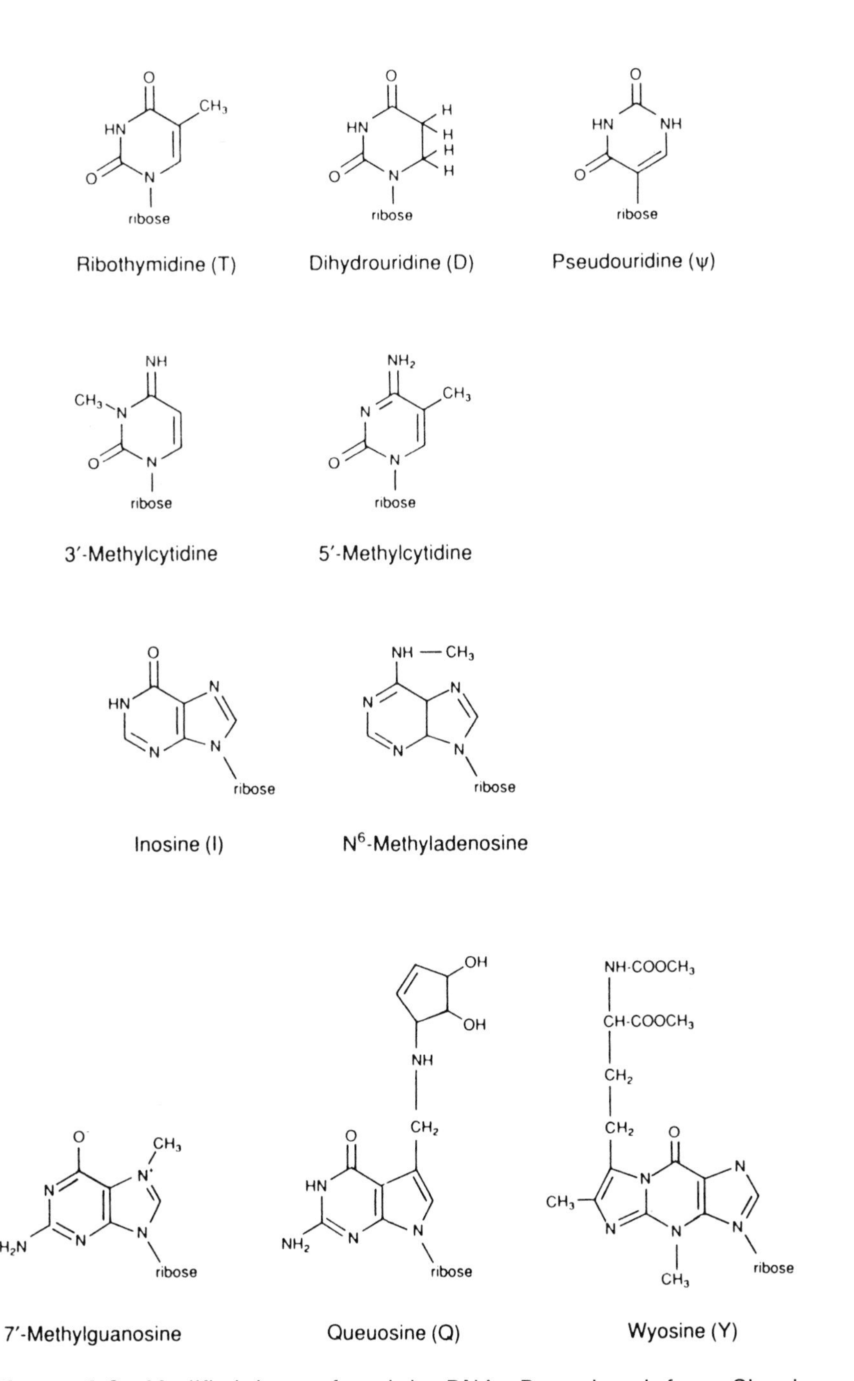

Figure 1.2: Modified bases found in RNA. Reproduced from Chambers and Rickwood (eds) (1993) *Biochemistry Labfax*, p. 249, BIOS Scientific Publishers Ltd.

frequently show unusual base pairing of G:U in addition to the expected Watson–Crick base pairing of A:U and G:C (*Figure 1.3*). Helix stability appears to require at least three conventional base pairs and the end loop of a hairpin appears always to have a minimum of three nucleotides. It is now possible to predict double-helical sections by computer analysis of primary sequence data. This technique has been used extensively to identify secondary structural components and homologies between a variety of RNA species.

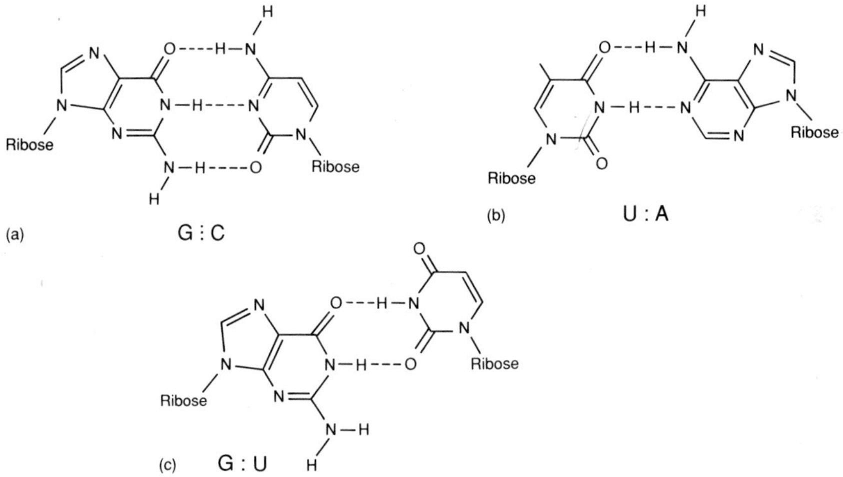

(a) G : C

(b) U : A

(c) G : U

Figure 1.3: Base pairing in RNA. (a), (b) and (c) show the structure of the types of base pairs found in helical secondary structures.

1.4 Diversity of types and functions of RNA

The three major classes of RNA in cells are messenger RNA (mRNA), transfer RNA (tRNA) and ribosomal RNA (rRNA); the last two types of RNA are often termed stable RNAs because they have a much longer half-life than mRNA. Of these three types of RNA, rRNA constitutes more than 80% of the total cell RNA. Prokaryotic cells contain three species of rRNA whereas cytoplasmic ribosomes of eukaryotic cells contain four species, but both sets of rRNAs confer similar properties to both prokaryotic and eukaryotic ribosomes. Most cells contain 60–70 different tRNAs, whereas the number of mRNAs exceeds several thousand and these can be of widely varying length. Mitochondria and chloroplasts have their own populations of mRNA, tRNA and rRNA. Besides the three major classes of RNA, there are also other RNAs with important cellular functions. Several RNAs, collectively called small nuclear RNAs (snRNAs), are found in the nuclei of eukaryotic cells and are involved in splicing mRNA precursors. Another RNA, designated 7S RNA, is an essential component of a cytoplasmic ribonucleoprotein (RNP) particle which is involved in the transport of proteins through phospholipid membranes. In addition, there are examples of specific RNA molecules which are essential components of enzymes. In many cases, the RNA alone has the catalytic activity. Thus, RNAs are a highly versatile class of molecules possessing a wide range of biological activities.

1.4.1 mRNA and hnRNA

mRNA molecules are now known to account for 3–5% of the total cellular RNA. They are quite heterogeneous with regard to size, reflecting the diversity of the gene products which they encode. They are synthesized on a DNA template by transcription. Base pairing is used to synthesize RNA complementary to the template strand of DNA.

Relatively little is known about the precise secondary and tertiary structure of mRNA. Most, but not all of the mRNA in living cells is found associated with ribosomes in the form of polyribosomes (polysomes). However, some cytoplasmic mRNAs are found in the cytosol as a ribonucleoprotein complex called messenger ribonucleoprotein (mRNP) and in some cases the mRNA in these complexes is masked, so that it is not translatable.

The most evident difference in RNA synthesis is that in eukaryotes the mRNA is synthesized as a large precursor molecule in the nucleus. This nuclear RNA is different from mRNA in that its average size is much larger, it is generally unstable, and it has a much greater sequence complexity. Taking its name from its broad size distribution, it is called heterogeneous nuclear RNA (hnRNA). It undergoes a complex process of maturation, usually involving a considerable reduction in size as well as other modifications. These include the addition of the so-called cap structure at the 5′ end and the poly(A) tail at the 3′ end, RNA editing, etc., before the mRNA is exported to the cytoplasm.

The instability of most prokaryotic mRNAs is striking. Most mRNAs in prokaryotes are enzymatically degraded within 1–3 min of their synthesis. On the other hand, the majority of eukaryotic mRNAs are stable for at least several hours. The degradation of these mRNAs is tightly controlled and thus this constitutes another level of regulation of gene expression.

1.4.2 tRNA

tRNA occupies a pivotal position in protein synthesis, since it is an adapter molecule that accomplishes the translation of each nucleotide triplet into an amino acid. The crucial feature that confers this capacity is the ability of tRNA to fold into a specific tertiary structure. The sequences of several hundred tRNAs from many sources have now been established and numerous studies including detailed X-ray crystallographic structural information have made it possible to define precisely the structure and function of tRNA. Almost all known tRNAs may be schematically arranged in the so-called clover-leaf secondary structure.

The clover-leaf form of the secondary structure of tRNA should not be taken too literally as a view of the tertiary structure, which actually is rather compact. The secondary structure folds into a compact L-shaped tertiary structure, with the anticodon at one extremity and the amino acid acceptor at the other. Many of the tertiary interactions are non-Watson–Crick associations. Moreover, most of the bases involved in these interactions are either invariant or semi-invariant, which strongly suggests that all tRNAs have similar conformations. The structure is also stabilized by several unusual hydrogen bonds between bases and either phosphate groups or the 2′-OH groups of ribose residues.

The set of tRNAs responding to the various codons for each amino acid is distinctive for each organism, and multiple tRNAs may respond to a particular codon. Of the 20 amino acids, each is recognized by a particular aminoacyl-tRNA synthetase, which also recognizes all the tRNAs coding for that amino acid. Aminoacyl-tRNA synthetases have a proofreading function which scrutinizes the aminoacyl-tRNA products and hydrolyzes any incorrectly charged tRNAs.

1.4.3 rRNA

The ribosome travels along the mRNA engaging in rapid cycles of peptide bond synthesis (see Section 1.5.1). It was originally viewed as a collection of proteins with various catalytic activities, held together by protein–protein interactions and binding to rRNA. However, the discovery of RNA molecules with catalytic activities immediately suggested that rRNA might play a more active role in ribosome function. There is now evidence that rRNA interacts with mRNA and/or tRNA at each stage of translation. Other extensive biochemical, genetic and phylogenetic evidence suggests that the rRNA may participate in mRNA selection, tRNA binding, ribosomal subunit association, proofreading, factor binding, antibiotic interactions, nonsense and frame-shift suppression, termination and the peptidyl transferase function [1].

A major function of the rRNAs is clearly structural. Proteins bind to each major rRNA at particular sites and in a specific order required to assemble each subunit. Indeed, ribosomes are interesting not only for their function, but also for the process by which they self-assemble from the constituent RNA and protein molecules.

1.4.4 snRNA

In the cell there is an important group of low-molecular-weight or small nuclear RNAs (snRNAs) which are metabolically quite stable. The snRNAs account for about 0.5% of the total cellular RNA and can usually be resolved into 11–14 well-defined species by gel electrophoresis. They range in size from 30 to 93 nucleotides and have a varied nucleotide composition and sequence. Five main snRNAs, U1, U2, U3, U4 and U5 have been identified. All U snRNAs involved in *cis*-splicing share two characteristic features: a 2,2,7-trimethylguanosine (TMG) cap structure and an Sm antigen-binding consensus sequence in the 3′ domain of the molecule. The U6 snRNA is an exception; it contains a monomethylguanosine structure rather than TMG.

In cell extracts, the snRNAs exist in association with proteins as small nuclear ribonucleoprotein complexes (snRNPs, pronounced 'snurps'). In an *in vitro* splicing reaction, the snRNPs, other essential protein factors [2], and the pre-mRNA form a macromolecular complex known as a spliceosome. It is still not clear how and where spliceosomes are formed and whether structures comparable to *in vitro* spliceosomes occur in the nucleus.

The snRNAs play a vital role in RNA processing, especially in RNA splicing and polyadenylation (see Section 1.6.3). Their functions largely depend upon their secondary and tertiary structures. For example, in RNA splicing, the 5′ splice site and branch site sequence are recognized in part by base pairing with the U1 and U2 RNAs, respectively. The interaction between the 5′ end of the U6 snRNA and 3′ end of U4 snRNA, while it is not required for pre-mRNA splicing, is important for snRNP assembly. Recently, experiments have shown that base pairing between U2 and U6 snRNAs is necessary for splicing a mammalian pre-mRNA [3,4]. The interaction of U2 and U6 results in two closely juxtaposed RNA helices and may reflect part of a higher-order RNA structure involving snRNAs and the pre-mRNA that is important for catalyzing splicing. U7 snRNA has also been shown to be involved in the cleavage/polyadenylation of histone mRNA in *Xenopus laevis*.

1.5 Ribonucleoproteins (RNPs)

1.5.1 Ribosomes and polysomes

Proteins are synthesized on ribosomes, these are large defined complexes of RNA and protein molecules. Eukaryotic and prokaryotic ribosomes are similar in design and function. Each is composed of one large and one small subunit, which fit together to form a complex with a mass of several million daltons (Da). The small subunit binds the mRNA, while the large subunit binds charged tRNA and catalyzes peptide bond formation.

All ribosomes in a bacterium are identical. In *E. coli*, the ribosomal RNAs have been sequenced and the amino acid sequences of the proteins have been determined. The small (30S; see Section 3.2.1) subunit consists of the 16S rRNA and 21 proteins. The large (50S) subunit contains the 23S rRNA, the small 5S RNA and 31 proteins. With the exception of one protein present at four copies per ribosome, there is one copy of each protein. About 60% of the mass of the particle is rRNA.

The cytoplasmic ribosomes of higher eukaryotic cytoplasm are larger and are more complex than those of bacteria. Eukaryotic ribosomes have particle masses in the range $3.9\text{–}4.5 \times 10^6$ Da and have a nominal sedimentation coefficient of 80S. They dissociate into two unequal subunits that have compositions that are distinctly different from those of prokaryotes. The small (40S) subunit of the rat liver cytoplasmic ribosome, the best-characterized eukaryotic ribosome, consists of 33 unique polypeptides and an 18S rRNA. The large (60S) subunit contains 49 different polypeptides and three rRNAs of 28S, 5.8S and 5S. RNA is still a major component by mass but comprises only about 50% of the total. Sequence comparisons of the corresponding rRNAs from various species indicate that evolution has conserved their secondary structures rather than their base sequences.

Ribosomes have three binding sites for RNA molecules: one for mRNA and two for tRNAs. One site, called the peptidyl-tRNA binding site, or P-site, holds the tRNA molecule that is linked to the nascent polypeptide chain. Another site, called the aminoacyl-tRNA binding site, or A-site, binds the incoming tRNA molecule charged with an amino acid. A tRNA molecule is bound tightly at either site only if its anticodon forms base pairs with adjacent codons in the mRNA molecule.

There is increasing evidence that the rRNA molecules play a central part in its catalytic activities. Ribosomal proteins, unlike rRNAs, have been relatively poorly conserved in sequence during evolution, and a surprising number of proteins do not seem to be essential for ribosome function. Therefore, it has been suggested that the ribosomal proteins mainly enhance the function of the rRNAs and that the RNA molecules rather than the protein molecules catalyze many of the ribosomal reactions. The ribosome represents a collection of many enzymes, each active only in the context of the overall structure, whose coordinated activities together accomplish the reactions involved in translation.

Many of the proteins and the rRNA elements may be concerned principally with establishing the overall structure that brings the various active sites into the right relationship and need not necessarily participate directly in the synthetic reactions.

Thus the ribosome provides an environment that controls the recognition between a codon of mRNA and the anticodon of tRNA. All the ribosomes of a given cell compartment appear to be identical. They undertake the synthesis of different proteins by associating with the different mRNAs that provide the actual templates. When fractions of active ribosomes are isolated they are found to be complexed along mRNA strands each of which bears many ribosomes; these complexes are known as polyribosomes or, more usually, polysomes.

1.5.2 hnRNPs and snRNPs

Newly synthesized RNA in eukaryotes, unlike that in bacteria, appears to become immediately condensed into a series of closely spaced protein-containing particles. These particles consist of about 500 nucleotides of RNA wrapped around a protein complex that serves to condense and package each growing RNA transcript. The resulting hnRNPs can be purified after nuclei have been treated with ribonucleases at levels just sufficient to destroy the linker RNA between particles. The particles sediment at about 30S and have a diameter of about twice that of nucleosomes (20 nm). Their protein core is correspondingly larger and more complex, being composed of a set of at least eight different proteins of mass 34–120 kDa. The proteins in this core are highly abundant in the cell nucleus. Those characterized thus far contain one or more copies of a short motif of amino acids that is shared by many RNA-binding proteins.

snRNPs each contain a set of proteins complexed to a stable RNA molecule. Individual snRNPs are believed to recognize specific nucleic acid sequences through RNA–RNA base pair complementarity. The snRNPs are much smaller than ribosomes, about 250 kDa compared with 4500 kDa for a ribosome – and have a higher protein-to-RNA ratio. Some proteins are present in several types of snRNPs, whereas others are unique to one type.

Previous studies have reported the existence of a reticular network of snRNPs within the nucleoplasm, extending between the nucleolar surface and the nuclear envelope [5]. Three snurposomes (A, B and C), consisting of different snRNA and protein components found in amphibian oocyte nuclei, suggests that different components of the pre-mRNA processing apparatus associate with nascent transcripts at different times and in different places within the nucleus.

1.5.3 mRNPs

Eukaryotic mRNA is usually complexed with a number of proteins. One major protein that is characteristic of all polyribosomal mRNPs is the 'poly(A)-binding protein' which has a mass of 73 kDa. Electron microscopy indicates extensive coverage of the mRNA and the proteins probably interact at precisely defined sites along the messenger molecule. Non-polysomal mRNPs differ substantially from the polysomal mRNPs and the poly(A)-binding protein is absent.

1.5.4 Ribozymes

The catalysis of reactions in biology has long been considered to be a purely protein characteristic. However, RNA has recently been shown to catalyze certain reactions. RNA enzymes are often called ribozymes. Active sites may form when short regions are held in a particular structure through secondary/tertiary interactions providing a surface of active groups that can act on another molecule. These active groups and the catalytic interaction can be fulfilled by base pairing and, to date, all RNA catalysis involves RNA substrates.

There are four types of RNA-catalyzed reactions in which the underlying chemistry is similar: the removal of a proton from a properly positioned oxygen triggering an attack by that oxygen on the electrophilic phosphorus atom in the phosphodiester bond that is to be broken. The common result is the transesterification or hydrolysis of specific phosphodiester bonds in RNA. There are only a few cellular catalysts composed primarily of RNA, whilst most others are found complexed to a protein, indicating that the presence of RNA in an enzyme is probably the consequence of an unusual evolutionary circumstance or constraint.

Although there are only four definite known types of RNA-catalyzed reactions, the view of what RNAs can catalyze is expanding. It has also been suggested that spliceosomes are ribozymal systems that evolved from primordial self-splicing RNAs. Furthermore, the RNA components of ribosomes may also have catalytic functions.

1.6 RNA processing

RNA processing is a major feature of eukaryotic RNA synthesis. Four main classes of processing events have been identified in eukaryotic RNA.

1.6.1 5′-End 2,2,7-trimethylguanosine (TMG) cap structure

Class II gene transcripts are modified at their 5′ end by the addition of a so-called 'cap' structure (*Figure 1.4*), which plays a major role in translation. The cap is an added 5′ terminal G, methylated on the 7-position of the base and linked to the initiator nucleotide by an unusual 5′–5′ triphosphate. Caps also occur on at least one class III gene transcript, the snRNA U6.

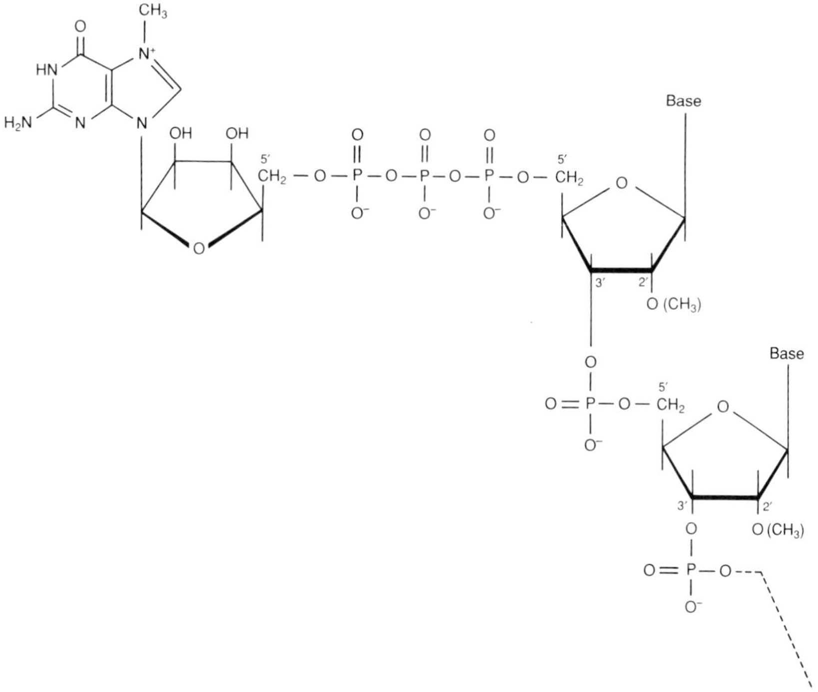

Figure 1.4: Cap structure of mRNA. Methyl groups in brackets indicate variable groups depending on the type of organism.

1.6.2 3′-End poly(A) tail addition

The majority of class II gene transcripts (with certain exceptions such as most histone mRNAs) contain a 3′ poly(A) tail, 40–100 residues long, which is added post-transcriptionally. The addition of this tail, by a special polymerase, is triggered by the signal AAUAAA encoded in the 3′ non-coding region of the mRNA; this signal is sufficient by itself in frog β-globin mRNA, but the human equivalent requires extra GU- and U-rich sequences 3′ to it for maximal efficiency.

1.6.3 RNA splicing

Transcripts of all three gene classes are usually subject to endonucleolytic or exonucleolytic cleavage followed, in the case of introns of class II genes, by exon splicing to form the functional mature species. The 5S RNA is an exception, being synthesized directly in the functional mature form. Class I genes are transcribed as a single large precursor from which the mature 18S, 5.8S and 28S rRNAs are derived via a well-characterized cleavage pathway. In class II gene transcripts with multiple introns there is no strict order in which introns are removed, but they tend to be removed from the 5′ end of the transcript first. Splicing proceeds in two steps (*Figure 1.5*). The first step is the cleavage of the intron at its 5′ consensus sequence, the free end of the intron folds round to form a lariat structure with the branch consensus box TACTAAC, itself located in the intron near the 3′ splice consensus sequence. In the second stage, the 5′-G in the intron at the 5′ splice junction then links to the 2′ hydroxyl group of ribose of the 3′-A residue in the RNA sequence UACUAAC. The 3′ splice site is then cleaved, the two exons ligated together and the excised intron is linearized and then degraded. This ATP-requiring reaction is probably carried out by a 50–60S 'spliceosome' complex containing proteins and snRNAs (see

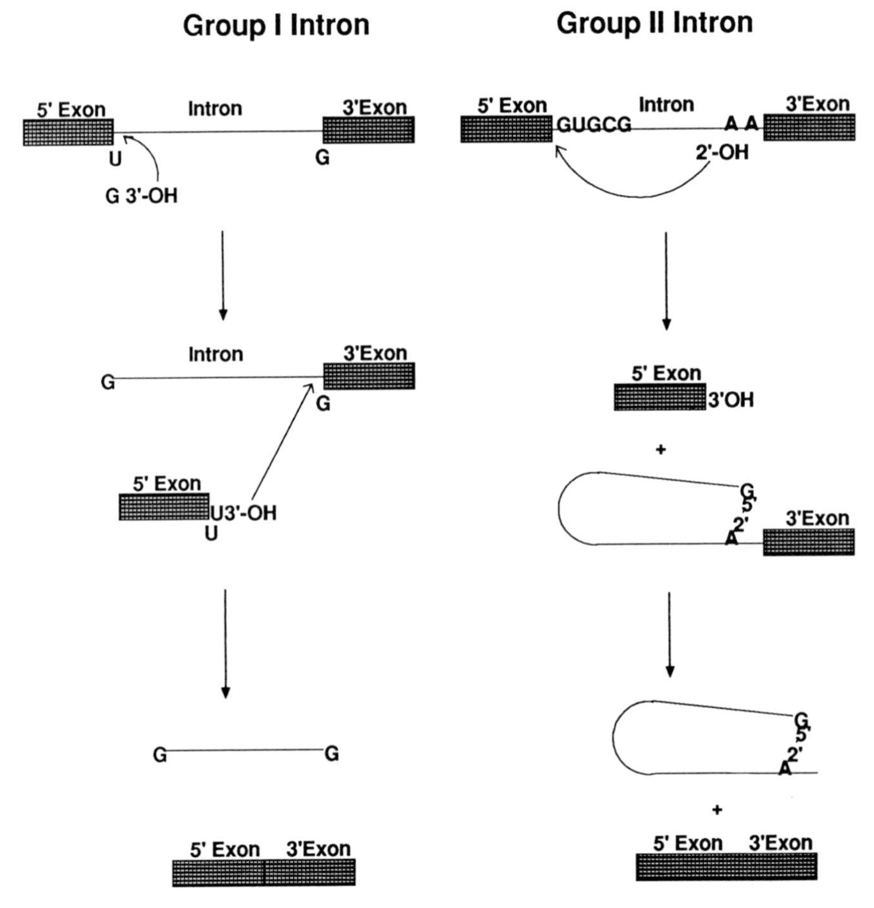

Figure 1.5: Mechanism of splicing of mRNA. The pathway of nuclear pre-mRNA splicing of Group I and II introns is shown schematically.

Section 1.4.4), such as U1 which have sequence complementarity to the 5′ exon–intron splice site.

The form of intramolecular intron removal and exon splicing is undoubtedly the most common pathway in eukaryotic nuclei. *Trans*-splicing has been observed in a number of cases, especially in trypanosomes. The *trans*-splicing reaction involves an RNA containing a 5′ splice site being joined to a separate (non-covalently linked) RNA containing a 3′ splice site. The *trans*-spliced RNA and protein-coding genes are under differential transcriptional controls. *Trans*-splicing has been reported in higher eukaryotes but its occurrence seems to be rather rare.

Some precursor transcripts give rise to multiple mature mRNAs because they can be processed in more than one way, that is, alternatively spliced. Alternative splicing increases the number of protein products of a single gene and provides an additional level of gene regulation.

Some primary transcripts can, in some cases, process themselves. Self-splicing *in vitro* is a property of the 26S rRNA from *Tetrahymena* and of some mitochondrial DNA transcripts from lower eukaryotes; intron removal *in vitro* requires only a guanine nucleotide, magnesium ions and monovalent cations. These RNA molecules are truly autocatalytic, requiring no proteins to assist in the splicing reaction.

1.6.4 mRNA editing

mRNA editing is the term coined to describe the post-transcriptional alteration of the informational content of mRNA independent of template-directed transcription. It suggests the correction of apparent coding errors (that is, frame-shifts or 'wrong' residues), although the modifications can involve insertion and deletion of hundreds of residues, including the creation of new reading frames [6]. In subtle forms of mRNA editing, specific nucleotides are modified to alter the mRNA sequence. Of the first six instances of RNA editing described [7], four involve mitochondrial systems, another a viral system and only one nuclear gene; RNA editing could have been considered an interesting phenomenon of limited relevance. However, two cases of editing subsequently found, relate to the editing of transcripts of chloroplasts [8], and the other at least three nuclear transcripts in the mammalian systems [9], further broadening the impact of RNA editing and suggesting that more unrecognized editing cases may exist.

Analogous to RNA splicing, differential editing is also observed. For several plant mitochondrial genes, comparisons of independent cDNA clones show that transcripts of the same gene may be edited to different degrees. If all these mRNAs were equally accessible for translation, different polypeptides could be synthesized as a result of each editing process [10].

References

1. **Noller, H.F.** (1991) *Annu. Rev. Biochem.*, **60,** 191.
2. **Fu, X.D. and Maniatis, T.** (1990) *Nature,* **343,** 347.
3. **Wu, J. and Manley, J.L.** (1991) *Nature,* **352,** 818.
4. **Datta, B. and Weiner, A.M.** (1991) *Nature,* **352,** 820.
5. **Spector, D.L.** (1990) *Proc. Natl Acad. Sci. USA,* **87,** 147.
6. **Stuart, K.** (1991) *Trends Biol. Sci.* **16,** 68.
7. **Cattaneo, R.** (1991) *Annu. Rev. Genet.,* **25,** 71.
8. **Hoch, B., Rainer, M.M., Appel, K., Gabor, L.I. and Kossel, H.** (1991) *Nature,* **353,** 178.
9. **Sommer, B., Kohler, M., Sprengel, R. and Seeburg, P.H.** (1991) *Cell,* **67,** 11.
10. **Sissinger, B., Brennicke, A. and Schuster, W.** (1992) *Trends Genet.,* **8,** 322.

Further reading

Arnstein, H.R.V. and Cox, R.A. (1992) *Protein Biosynthesis in Focus* (D. Rickwood, ed.). IRL Press, Oxford, UK.

Beebee, T. and Burke, J. (1992) *Gene Structure and Transcription in Focus,* 2nd edn. IRL Press, Oxford, UK.

Lamond, A. (1991) Nuclear RNA processing. *Curr. Opin. Cell Biol.,* **3,** 493.

Mattaj, I.W. (1990) Splicing stories and poly(A) tales: an update on RNA processing and transport. *Curr. Opin. Cell Biol.,* **2,** 528.

Moore, P.B. (1992) RNA catalysis: the universe expands. *Nature,* **357,** 439.

Pace, N.R. and Smith, D. (1990) Ribonuclease P: function and variation. *J. Biol. Chem.,* **265,** 3587.

Steitz, J.A. (1992) Splicing takes a Holliday. *Science,* **257,** 888.

2 Methods for isolating RNA

2.1 Basic principles

One of the most important aspects in the isolation of RNA is to prevent any degradation of the RNA during the isolation procedure. Since RNA breakdown is a universal feature of all cells during the processing and metabolism of RNA, all cells possess a wide range of nucleases with both specific and non-specific activities. Both nuclei and lysosomes are particularly rich in nucleases but nuclease activity is not restricted to these organelles. Some tissues, such as pancreatic tissue, have particularly high nuclease activity. Given the wide distribution of nucleases and the potential damage that they can inflict on RNA during isolation, it is important to ensure that nuclease activity is totally inhibited, a problem made all the more difficult by the robust nature of many nucleases.

A few golden rules can be given and it is important that they are followed if the isolation procedure is to be successful; all too often the first attempts to isolate RNA leave the novice with only degraded fragments of RNA. Firstly, it is important to ensure that all the reagents used are as free as possible of nucleases. All solutions should be made from either fresh material or material whose history is known using autoclaved double-distilled water and, where possible, autoclaving the final solution. Solutions used in the initial extraction (during which nucleases are released) are often supplemented with nuclease inhibitors such as 0.1% 8-hydroxyquinoline (*Table 2.1*). A pH range of 6.8–7.2 for solutions is normally a necessity to ensure that no RNA degradation occurs, as RNA is degraded at an alkaline pH as low as 9. Glassware should be oven-baked at 200°C for 4–12 h, whereas plastic is usually autoclaved. Skin is a well-known source of nucleases and so it is essential to wear disposable plastic or latex gloves during all manipulations.

Cellular RNA is, for the most part, complexed with proteins some of which have nuclease activity. Hence it is important that any isolation method ensures rapid inactivation of cellular proteins. This can be achieved by using nuclease inhibitors such as hydroxyquinoline but

Table 2.1: Agents used to preserve RNA integrity during extraction

Agent	Active concentration	Method of ribonuclease inactivation	Comments
Aurintricarboxylic acid (ATA)	10 μM	Complexes to a wide range of nucleases	
Bentonite	3 mg/ml	Inactivation by adsorbing to nucleases	Purified suspension of clay particles
Diethylpyrocarbonate (DEPC)	0.1%	Alkylates proteins disrupting protein structure	Toxic, can modify bases of RNA
EDTA	1–10 mM	Chelates divalent cations needed for activity of some ribonucleases	Not all nucleases need divalent ions for activity
Guanidine hydrochloride	8 M	Denatures proteins	Toxic
Guanidium thiocyanate	4 M	Denatures proteins	Toxic, strong agent for ribonuclease inactivation
Heparin	0.5 mg/ml	Binds to basic ribonucleases	Can strip proteins from DNA
8-Hydroxyquinoline[a]	0.1% (w/v)	Inactivates ribonucleases	Toxic
Macaloid	0.015% (w/v)	Adsorbs ribonucleases	
Phenol/chloroform	50% (v/v)	Extracts and denatures ribonucleases	Toxic
Placental RNase inhibitor	1 U/μl	Protein inhibitor of RNases	RNasin™—Promega Inhibit-Ace™—5 prime–3 prime, Inc RNAguard™—Pharmacia
Polyvinyl sulfate (PVS)	1–10 μg/ml	Complexes to basic nucleases	
Proteinase K	100–200 μg/ml	Hydrolysis of proteins	Predigest to remove any nuclease contamination
Sodium dodecyl sulfate (SDS)	0.1–1%	Disrupts protein structures	Hazardous in powder form
Ribonucleoside vanadyl complex	10 mM	Binds to active site of ribonuclease	Reversible inhibition

[a]8-Hydroxyquinoline/8-hydroxy-1-azanaphthalene.

usually better results can be obtained by denaturing the proteins in the sample using strong detergents, chaotropic solutions or other protein denaturants (*Table 2.1*). In addition, it is a good rule to minimize any fractionation of the source material before starting the RNA isolation procedure. For example, nuclear RNA sequences can be degraded during the purification of nuclei unless great care is taken.

Extraction procedures are numerous due to different research requirements. The choice of procedure is usually based upon the tissue source, the type of RNA to be extracted, the cost of the procedure and the analytical resolution needed.

2.2 Cell lysis and fractionation procedures

The initial stages of extraction must, of necessity, lyse the cells. The severity of this treatment is dependent on whether a cell wall is present and its nature; it is important to avoid procedures which may allow or cause degradation of the RNA.

2.2.1 Isolation of total RNA by cell lysis

Bacteria. These are grown in broth cultures before isolation and then centrifuged at 5000 *g* for 10 min to pellet the cells. Cells are suspended in 0.15 M NaCl, 10 mM Tris-HCl (pH 8.0), 1 mM EDTA and lysozyme is added to a final concentration of 0.1 mg/ml. The cells are incubated at room temperature for 30 min to digest the bacterial cell wall before the addition of SDS to a final concentration of 1%. The RNA can then be extracted using the hot phenol procedure (see Section 2.3).

Animal cells. These can be taken from tissue culture or directly from a donor animal. The former source allows easy collection of cells, normally by scraping a rubber policeman over the sheet-like layers in the culture dishes, pouring media off and lysing the cells directly. Tissue taken directly from an animal is often solid and so it must be broken up as well as lysed. The usual method of breaking tissue apart is homogenization. Homogenization can be achieved by using a Potter–Elvejhem or Dounce homogenizer (*Figure 2.1a*) for soft tissues such as liver or kidney. Homogenization using a Polytron or Waring blender (*Figure 2.1b*) is more efficient for tough tissues such as adult muscle. An alternative, but seldom used, approach is to use proteolytic enzymes, such as collagenase, before homogenization.

Any mechanical disruption of animal cells must be carried out in the presence of a homogenization buffer which will ensure that the RNA is not degraded before the preparation has started. Cell lysis solutions must disrupt cell membranes as well as inhibit nucleases, while avoiding any chemical degradation of RNA. This is often achieved by use of a solution containing urea, SDS or strong chaotropic agents. *Table 2.2* details some lysis buffers and tissues for which they can be used.

Plant and fungal cells. Like bacteria, plant and fungal cells have strong, rigid cell walls surrounding their cell membrane which must be broken, either by strong mechanical action or by enzymatic digestion (protoplasting).

For mechanical breakage, fresh or frozen tissue can be ground in prechilled mortars in a pool of liquid nitrogen until a fine powder is achieved, before transferring to a deproteinization solution. It is

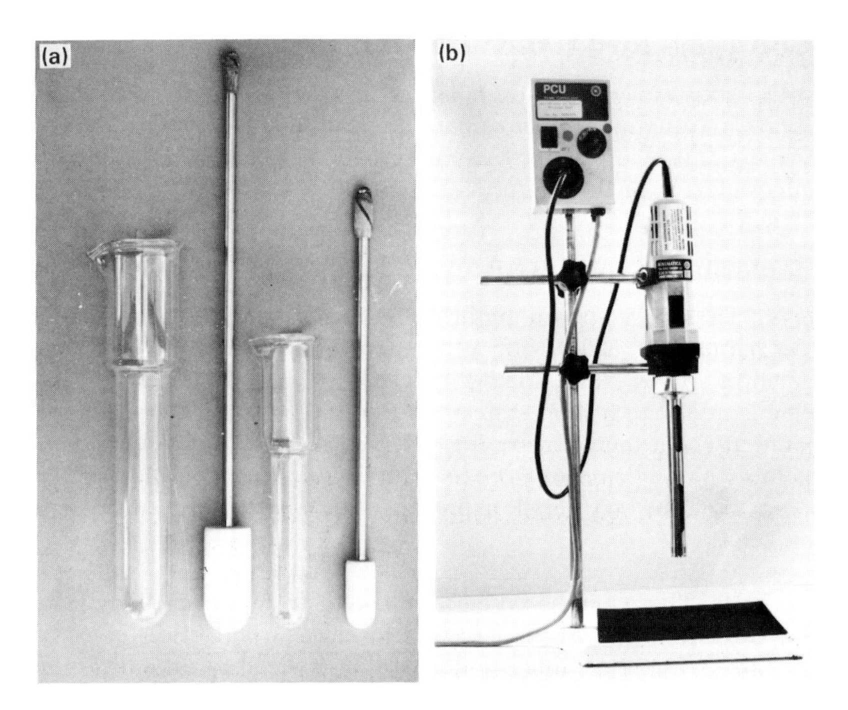

Figure 2.1: Homogenizers and blenders for disrupting tissues and cells. (a) Potter-Elvejhem homogenizer with a Teflon pestle and glass outer vessel; (b) Polytron blender.

very important that eye protection is worn for this procedure. Further cell lysis can be accomplished by high-speed shearing with a Polytron blender.

For protoplasting plant cells the fresh tissue is finely sliced using a scalpel and transferred to digestion medium. For each gram of tissue use 25 ml of 0.6 M mannitol, 10 mM MES-NaOH (pH 5.2), 1 mM $CaCl_2$, 1 mM $MgCl_2$, 10 mM 2-mercaptoethanol containing 2 mg/ml bovine serum albumin (BSA), 2 mg/ml 'macerase' (pectinase–Cambio), 20 mg/ml 'cellulysin' (cellulase–Cambio). The mixture is incubated at 25°C for 3 h. Protoplasts may be collected by centrifugation at 1000 *g* for 5 min at 20°C and washed in isotonic medium. The protoplasts can then be lysed using the same methods as are used for animal cells.

In the case of fungal cells a similar protoplasting procedure is used though the choice of enzymes tends to vary depending on the type of fungal cell. For example, for yeast, a crude extract of snail gut enzyme is often used.

To subfractionate the cells before extracting the RNA, for example in the isolation of nuclei for the preparation of nuclear RNA, protoplasting is often the best way of obtaining subcellular fractions of acceptable purity.

2.2.2 Isolation of RNA from cell fractions

Extraction procedures can have general targets (extracting total cellular RNA) or specific targets (cytoplasmic, nuclear or organellar RNA). Extraction of total cellular RNA requires no specific separation of cellular organelles after lysis or homogenization, allowing direct application of an extraction procedure without any prefractionation of the cells.

Cytoplasmic and nuclear RNA. This RNA can be isolated by homogenization of the cells in a non-disruptive buffer (e.g. buffer 1 of *Table 2.2*) followed by pelleting of the nuclei using Protocol 2.1.

Note that the nuclear pellet obtained in step 2 of Protocol 2.1 is contaminated with significant amounts of other types of intracellular material especially large mitochondria and large sheets of endoplasmic reticulum both of which contain RNA that is distinct from nuclear RNA.

Table 2.2: Lysis buffers used in RNA extraction procedures

Buffer	Constituents	Disruption agent	RNase inactivation agent	Comments
1	10 mM Tris-HCl (pH 8.6), 0.14 M NaCl, 1.5 mM $MgCl_2$, 0.5% NP40[a], 1mM DTT[b], 20 mM vanadyl ribonucleoside complexes	NP40	Vanadyl ribonucleoside complexes	Used in cytoplasmic and total RNA preparations of mammalian tissue culture cells
2	10 mM EDTA (pH 8.0), 0.25% (w/v) SDS[c], 50 mM sodium acetate (pH 5.2)	SDS	SDS and EDTA	Used in rapid preparations of mammalian total RNA
3	50 mM Tris-HCl (pH 7.5), 50 mM NaCl, 5 mM EDTA (pH 8.0), 0.5% SDS, 200 μg/ml proteinase K	Homogenization	Proteinase K, SDS and EDTA	Homogenization buffer uses mechanical lysis as well as some chemical lysis
4	50 mM Tris-HCl (pH 9.0), 100 mM NaCl, 10 mM EDTA (pH 8.0), 2% SDS, 100 μg/ml proteinase K	Homogenization	Proteinase K, SDS and EDTA	Homogenization buffer uses mechanical lysis as well as some chemical lysis. Used in the phenol extraction system on plant tissue
5	150 mM Tris-HCl (pH 8.5), 0.1 M NaCl, 1% SDS, 20 mM EDTA, 0.5 mg/ml heparin	Homogenization	Heparin, SDS and EDTA	Used in mRNA stability assessments

LIVERPOOL JOHN MOORES UNIVERSITY AVRIL ROBARTS LRC TEL. 0151 231 4022

Table 2.2: Continued

Buffer	Constituents	Disruption agent	RNase inactivation agent	Comments
6	8 M guanidine-HCl, 0.1 M sodium acetate (pH 5.2), 5 mM DTT, 0.5% sodium lauryl sarcosinate	Homogenization	Guanidine-HCl	Guanidine-HCl method has largely replaced the phenol chloroform extraction method
7	8 M guanidine-HCl, 20 mM MES[d], 20 mM EDTA (pH 8.0), 0.22 M 2-mercaptoethanol and adjusted to pH 7.0 with NaOH	Homogenization	Guanidine-HCl and EDTA	Used for rapid total RNA preparations
8	4.0 M guanidinium thiocyanate, 0.1 M Tris-HCl (pH 7.5), 0.14 M 2-mercaptoethanol	Homogenization	Guanidinium thiocyanate	Guanidinium thiocyanate is capable of preventing RNA degradation in disrupted tissue that possesses high endogenous levels of RNases
9	4.0 M guanidinium thiocyanate, 25 mM sodium citrate (pH 7.0), 0.5% sarcosyl, 0.1 M 2-mercaptoethanol	Homogenization	Guanidinium thiocyanate	Guanidinium thiocyanate is capable of preventing RNA degradation in disrupted tissue that possesses high endogenous levels of RNases
10	RES-1: 1 M urea, 0.5 M LiCl, 0.02 M sodium citrate, 2.5 mM cyclohexanediaminetetracetate (CDTA) (pH 6.8) and 1% (w/v) SDS	Homogenization	Urea and SDS	
11	0.15 M sucrose, 10 mM sodium acetate, 1% (w/v) SDS	Lysis	SDS	Used in the hot phenol method and for bacterial RNA isolation
12	GTEM: 5.0 M guanidinium isothiocyanate, 50 mM Tris-HCl (pH 7.5), 10 mM EDTA, 1.12 M 2-mercaptoethanol	Lysis	Guanidinium thiocyanate	

[a]NP40—Nonidet P-40, detergent; [b]DTT, dithiothreitol; [c]SDS, sodium dodecyl sulfate; [d]MES, 2-(*N*-morpholino)ethanesulfonic acid.

Protocol 2.1: Nuclei isolation

1. Large cell debris and unbroken cells are removed by centrifugation at 500 *g* for 3 min at 4°C.
2. Nuclei are pelleted from the homogenate by centrifugation at 2500 *g* for 10 min at 4°C.
3. SDS and NaCl are added to the post-nuclear supernatant from step 2 to final concentrations of 1% (w/v) and 0.3 M, respectively.
4. RNA can then be isolated by phenol extraction (see Section 2.3).

Nuclei can be purified by sedimentation through dense (2.1 M) sucrose or by using an ice-cold citric acid protocol the latter of which has the advantage of reducing nuclease activity by virtue of the low pH of the citric acid solutions used. If the cytosolic fraction is not required then nuclei can be purified directly from cells and tissues using the citric acid method (Protocol 2.2).

Protocol 2.2: Nuclei isolation using the citric acid method

1. Pellets of tissue culture cells, finely chopped tissue or crude nuclear pellets are suspended in ice-cold 5% (w/v) citric acid (1 ml/10^8 cells) by vigorous vortexing.
2. The suspension is homogenized with 10–15 strokes of a tight-fitting Teflon pestle in a glass Potter homogenizer vessel immersed in ice.
3. Centrifugation at 250 *g* for 5 min at 4°C loosely pellets the nuclei. Care must be taken when aspirating the supernatant to ensure no loss of the nuclear pellet occurs.
4. Pellets are resuspended in ice-cold 0.25 M sucrose in 1.5% (w/v) citric acid (1 ml/10^8 cells) by vigorous vortexing.
5. The suspension is homogenized with five strokes of a tight-fitting Teflon pestle in a glass Potter homogenizer vessel immersed in ice.
6. The suspension is underlayered with ice-cold 0.88 M sucrose in 1.5% (w/v) citric acid and centrifuged at 1000 *g* for 5 min at 4°C.
7. The supernatant is discarded by aspiration with care being taken to aspirate the material at the interface before the lower layer.
8. Steps 4–7 are repeated, monitoring the nuclei purity with phase-contrast microscopy.
9. Step 8 is repeated without homogenization until the supernatant is clear and the nuclear suspension is free of debris.

RNA from cell organelles. When RNA is to be prepared from cell organelles, cell membrane lysis must leave the organelles intact. This is achieved by lysis treatments where the majority of the tissue is mechanically or chemically disrupted in a suitable isotonic medium without breaking the organelle, followed by centrifugation to pellet cell debris. Care must be taken to minimize cross-contamination of organelles such as mitochondria and chloroplasts with nuclear material as a result of disruption of the nuclei.

Whenever possible, it is best to purify organelles by isopycnic gradient centrifugation rather than using the crude pellets obtained by simple differential pelleting; gradients of Percoll, Nycodenz and sucrose have all been used for purifying particular organelles from a wide range of organisms. In choosing a gradient system for purifying organelles the best advice is to follow a published method that is widely accepted. If there has been disruption of the nuclei it is possible that nuclear RNA and DNA will stick to the outside of the mitochondria. Mitochondria can be isolated from cells by freeze–thawing cells which disrupts the outer membrane [1] or the outer membrane can be removed using digitonin [2].

Even when all possible precautions are taken, it is still possible for the RNA to be degraded; indeed RNA processing often starts before synthesis of the molecule is completed. Detailed analysis of the RNA is necessary, before one can be confident that the RNA isolated is indeed the original transcript, particularly in systems where RNA processing is extensive. It is good routine practice to determine the size and integrity of the isolated RNA using electrophoresis (see Section 3.3). In the case of cytoplasmic or total cellular RNA, the two bands of rRNA which constitute the majority of the cellular RNA should be clearly visible and running with the correct mobility (*Figure 2.2*), if they appear to be of much smaller molecular weight than expected (i.e. have a higher mobility on the gel, creating a smear on the gel) then it is likely that the RNA preparation is extensively degraded.

All RNA isolation methods include three steps: (i) the inactivation of nucleases, (ii) the dissociation of RNA from proteins, and (iii) the separation of the RNA from other macromolecules. Usually it is best if the first two steps are immediate and concurrent so that degradation of RNA is minimal. Total RNA may then be fractionated to give the particular type of RNA required; for example, mRNA or rRNA. Irrespective of the type of RNA required, a good general rule is to use as few steps as possible for purification procedures.

The remaining sections of this chapter describe the main methods for the isolation of total cellular RNA using procedures based on either chemical extraction or centrifugation techniques. In addition, methods for the isolation of mRNA, newly synthesized RNA using affinity

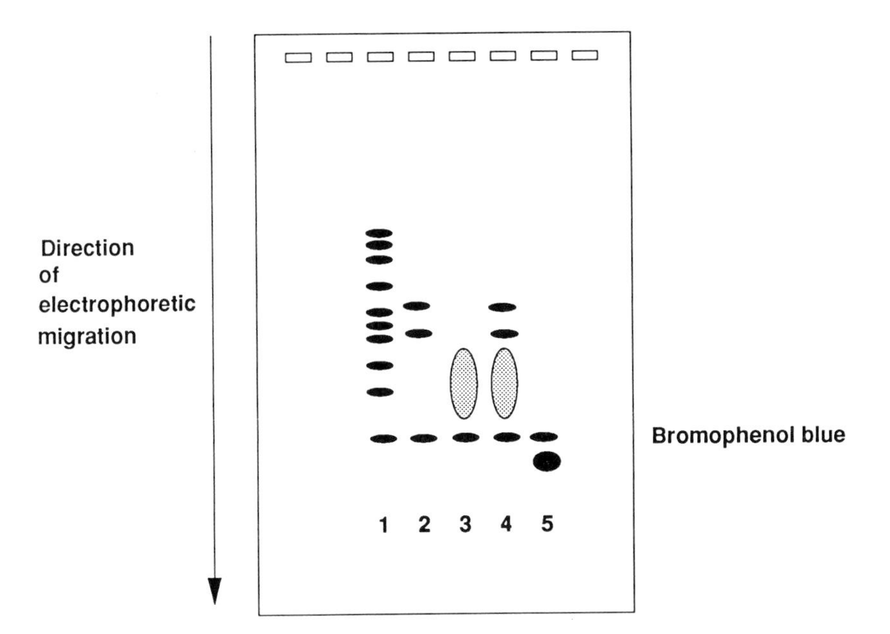

Figure 2.2: A schematic representation of the assessment of RNA integrity using gel electrophoresis. (1) Molecular size markers; (2) rRNA; (3) mRNA; (4) total RNA; (5) degraded RNA.

columns or magnetic beads, and small RNA molecules using high-pressure liquid chromatography (HPLC) are described.

2.3 Isolation of RNA using phenol extraction methods

This procedure was one of the first published methods to prepare nucleic acids free of contaminating proteins [3]. The procedure is based on the ability of organic molecules to denature and precipitate proteins in solution while not affecting the solubility of the nucleic acids. A number of procedures have been developed based around phenol containing various additives to improve the inhibition of nuclease activity (e.g. 8-hydroxyquinoline) or to improve the deproteinization efficiency of phenol (e.g. chloroform).

Phenol chloroform extraction is carried out by addition of an equal volume of buffer-saturated phenol–chloroform–isoamyl alcohol (25:24:1) to the sample, vigorous mixing to form an emulsion and separation of the denser phenol from the aqueous layer by centrifugation. High-purity phenol is available commercially but it is also possible to redistil the phenol before use, although this procedure can be hazardous. Phenol–chloroform–isoamyl alcohol (25:24:1) is prepared by equilibrating the pH of the acidic phenol–chloroform–isoamyl alcohol solution with an autoclaved buffer of the required pH (e.g.

0.1 M Tris buffer pH 7.0). An equal volume of buffer is mixed into the phenol–chloroform–isoamyl alcohol by vigorous shaking. The mixture is allowed to settle and the upper layer (aqueous buffer) removed. Equilibration with buffer is repeated until the required pH is reached.

The basis of the method (Protocol 2.3) is to suspend the cells in a solution (such as Buffer 4 – *Table 2.2*) containing a detergent which lyses the cells and dissociates nucleoprotein complexes.

Protocol 2.3: Isolation of RNA using the phenol method

1. Homogenization is accomplished in two 5 sec homogenizations with a Polytron homogenizer. The homogenization procedure varies with the type of tissue. Ground plant tissue is subject to two high-speed treatments, each for 5 sec [4]. Tough tissues such as pancreas have been subject to 30–60 sec of homogenization at high speed.
2. Digestion of protein over a period of 15 min at room temperature with 0.1 mg/ml proteinase K (a component of the homogenization buffer) removes much of the potential RNase activity.
3. Proteins are then removed by phenol–chloroform extraction. The extraction is performed by addition of an equal volume of buffer-saturated phenol–chloroform–isoamyl alcohol (25:24:1) and mixing well. Procedures often involve shaking the mixtures either by hand or in orbital incubators for 5–10 min to ensure the formation of a fine emulsion. The phenol–chloroform–isoamyl alcohol denatures and precipitates proteins so that when the emulsion is centrifuged (10 000 *g* for 10 min), the phenol–chloroform separates out as a dense bottom layer with the proteins either dissolved in the phenol–chloroform layer or precipitated at the interface, leaving the RNA in the top aqueous layer. Isoamyl alcohol is included to ensure that a well-defined interface is produced, making recovery of the supernatant easier.
4. The phenol extraction of the recovered aqueous phase is repeated with equal volumes of buffer-saturated phenol–chloroform–isoamyl alcohol (25:24:1) and the same shaking procedures until there is no protein at the interface after centrifugation. Usually three extractions are sufficient to yield a clear interface.
5. RNA in the aqueous phase is precipitated by the addition of either 2 vol. ethanol or 1 vol. isopropanol after the addition of sodium acetate (pH 5.2) to a final concentration of 0.3 M. After the addition of ethanol or isopropanol the RNA is left to precipitate at −20°C for at least 1 h.

6. RNA is pelleted by centrifugation at 10 000 *g* for 20 min at 0°C, but if only submicrogram amounts of RNA are being precipitated it is better to use a higher centrifugal force (e.g. 60 000 *g*) to ensure a high recovery of RNA.
7. After centrifugation the aqueous fraction is removed by aspiration and the pellet is washed with ice-cold 70% (v/v) ethanol to remove excess salt. The RNA pellet is then redissolved in approximately 500 μl sterile, double-distilled water (treated with DEPC) per 5 g starting material and transferred to a sterile 0.5 ml microcentrifuge tube.
8. The redissolved RNA is then mixed with an equal volume of phenol–chloroform–isoamyl alcohol to remove any residual protein. Centrifugation in a bench-top centrifuge for 3–5 min at 10 000–14 000 *g* at room temperature separates the phenol and aqueous layers.
9. The aqueous layer is recovered and extracted in an equal volume of chloroform–isoamyl alcohol at room temperature (to remove traces of phenol).
10. The RNA is then precipitated with 2 vol. ethanol and 1/10 vol. 3 M sodium acetate (pH 5.2). Precipitation for at least 2 h at −20°C precedes centrifugation at 4°C in a bench-top centrifuge at 10 000–14 000 *g* for 10–20 min. The RNA must then be redissolved in a suitable solution. Solutions commonly used to redissolve RNA can be found in *Table 2.3.*

CAUTION: Phenol is hazardous. It is a suspected carcinogen and should be used in fume cupboards to avoid inhaling vapors. Phenol burns skin badly, especially when hot. If burns occur in spite of protection the burnt area should be washed with 20% (w/v) polyethylene glycol (PEG) in 50% (v/v) industrial methylated spirits (IMS) *not* water. At least a liter of this solution should be available at all times. The use of latex gloves both for personal safety and to avoid degradation of the RNA is imperative.

This basic phenol extraction procedure has been modified by the inclusion of additives such as 8-hydroxyquinoline or *m*-cresol to phenol–chloroform–isoamyl alcohol to enhance RNA extraction [1]. Other approaches have been to modify the temperature of phenol extraction to 60°C ('hot phenol extraction') making the process more efficient, particularly for nuclear and bacterial RNA. An extension of this has been to carry out the series of phenol extractions at different temperatures and to obtain different fractions of RNA [5], but this method is hardly used now.

Table 2.3: Solutions used to dissolve RNA after extraction

Solution	Nature	Further procedures
Water (DEPC treated)	Non-denaturing	Agarose gel electrophoresis Preparative centrifugation Further molecular techniques
10% sucrose in formaldehyde	Denaturing	Denaturing polyacrylamide gel electrophoresis (PAGE)
10% sucrose in 8 M urea	Denaturing	Denaturing PAGE

The volume of solution used to redissolve the RNA varies with the amount of RNA isolated; as a general rule it is important that the RNA solution is not too dilute. The amount of RNA is usually directly related to the amount and type of the tissue that was initially extracted. Tissue types in which the yield using a given procedure is known, can be accurately assessed for the final volumes to dissolve the RNA in. For instance, cultured cells tend to yield about 25 µg total RNA/10^6 cells which can be redissolved in 100 µl sterile, double-distilled water. If RNA is difficult to dissolve in water it is likely that there is a high level of protein contamination and in this case an additional deproteinization step should be carried out.

The future use of the RNA also plays a role in the volume used to redissolve the RNA in. Gel electrophoresis has a maximum sample volume capacity defined by the size of the well and if rare RNA species are to be detected (Northern blotting, see Section 3.5.2) then the RNA must be dissolved in a small amount of solution. If the RNA is to be used to create a cDNA library it must be highly concentrated (5 µg of mRNA in 10 µl of water) and an RNA preparation from a large amount of tissue is therefore dissolved in approximately 20 µl.

A serious problem with phenol extraction is that, depending on the nature of the starting material, the RNA can be contaminated with significant amounts of both DNA and polysaccharides as both can be present in the aqueous layer after phenol extraction. This can be a serious problem because both contaminants are precipitated along with the RNA by ethanol or isopropanol and, depending on the subsequent use of the RNA, they may affect further analyses. DNA and polysaccharides can be removed from the RNA by enzymic degradation either by including a digestion step before the last phenol–chloroform–isoamyl alcohol extraction of the protocol or after extraction. DNA can be removed by addition of DNase I to a final concentration of 2 µg/ml and incubated at 37°C for 1 h.

It is important that any enzymes used are chromatographically purified before use to remove RNases. After enzymatic digestion of contaminating material, the RNA must be deproteinized again to remove these enzymes. This is accomplished by re-extraction with phenol–chloroform–isoamyl alcohol and re-precipitation with ethanol or isopropanol before it is used for further preparative analytical procedures. An alternative method to purify the phenol-extracted RNA is to pellet it through a cushion of 5.7 M CsCl, 0.01 M EDTA (pH 7.5) in an ultracentrifuge. This procedure will remove both DNA and polysaccharides as they are not dense enough to pellet through 5.7 M CsCl. However, if this latter approach is being considered it is probably better to use one of the centrifugation methods for the isolation of RNA described in Section 2.8.

2.4 Isolation of total RNA by acid guanidinium thiocyanate methods

The phenol–chloroform–isoamyl alcohol method has been largely superseded by the acid guanidinium thiocyanate method [6]. This method uses the chaotropic agent guanidium thiocyanate in which the cation and the anion are both strongly chaotropic. This allows a very fast inactivation of cellular RNases upon cell disruption. A fast inactivation is often essential making this the method of choice. The tissue types used with this method are highly diverse; bacterial plant and animal tissue have all successfully had RNA extracted from them (Protocol 2.4).

Protocol 2.4: Isolation of RNA using the acid guanidinium thiocyanate method

1. Cells are homogenized directly in a solution of 4 M guanidine thiocyanate containing 2-mercaptoethanol and 0.5% *N*-lauryl sarcosine pH 7.0 (*Table 2.2*).
2. The proteins are then separated from the nucleic acids by shaking the detergent-treated solution of proteins and nucleic acids with an equal volume of buffer-saturated phenol–chloroform–isoamyl alcohol (25:24:1) to form an emulsion.
3. As described previously, the phenol–chloroform–isoamyl alcohol denatures and precipitates the proteins so that when the emulsion is centrifuged (usually 10 000 *g* for 20 min at room temperature) the phenol–chloroform separates out as a dense bottom layer with precipitated proteins together with most of the DNA at the interface and the RNA in the top aqueous layer.
4. The upper aqueous layer is recovered with wide-bore disposable plastic Pasteur pipettes and the phenol extraction is repeated by adding equal volumes of phenol–chloroform–isoamyl alcohol to the recovered aqueous layer until there is no protein at the interface. In most cases three or four phenol–chloroform–isoamyl alcohol extractions are sufficient to remove all of the protein.
5. The RNA in the aqueous phase is precipitated by the addition of either 2 vol. ethanol or 1 vol. isopropanol after adding sodium acetate (pH 5.2) to a final concentration of 0.3 M.

LIVERPOOL
JOHN MOORES UNIVERSITY
AVRIL ROBARTS LRC
TEL 0151 231 4022

6. After the addition of ethanol or isopropanol, the precipitated RNA is left to aggregate at −20°C for at least an hour. The RNA can usually be pelleted by centrifugation at 10 000 *g* for 20 min at 0°C but if precipitating only submicrogram amounts of RNA it is better to use a higher centrifugal force (e.g. 60 000 *g* for 1 h) in order to ensure a high yield of RNA.
7. The RNA is re-dissolved in approximately 500 μl autoclaved, double-distilled water per 5 g starting material.
8. The remaining proteins are removed by a phenol–chloroform–isoamyl alcohol extraction using an equal volume of phenol–chloroform–isoamyl alcohol. The samples are shaken followed by centrifugation in a bench-top centrifuge at 10 000–14 000 *g* for 5 min.
9. A chloroform extraction of the recovered aqueous layer with an equal volume of chloroform–isoamyl alcohol removes any residual phenol. The samples are shaken followed by centrifugation in a bench-top centrifuge at 10 000–14 000 *g* for 5 min.
10. The RNA in the recovered supernatant is reprecipitated with 2 vols ethanol and 1/10 vol. 3 M sodium acetate (pH 5.2) at −20°C for at least 2 h. The RNA is collected by centrifugation at 10 000 *g* for 10 min at 0°C. The RNA is then dissolved in the required volume of a suitable solution as detailed in Section 2.3.

2.5 Isolation of total cellular RNA using the guanidinium–lithium chloride method

The use of lithium chloride in isolation methods is based on its ability to precipitate high-molecular-weight RNA species selectively [5]. This method has been adapted to extract RNA from small amounts of tissue (miniprep methods) to large amounts of tissue (3×10^6 mammalian cells/ml lysis solution). The type of tissue used in this procedure is unrestricted as it utilizes guanidium thiocyanate which can prevent RNase activity in most tissues and LiCl can precipitate all types of large RNA species. The method is, however, usually applied to mammalian cells which often have a high glycoprotein concentration that cannot be removed using standard phenol–chloroform–isoamyl alcohol extractions (Protocol 2.5).

Protocol 2.5: Isolation of RNA using the lithium chloride method

1. Cells are lysed in GTEM buffer (5.0 M guanidinium thiocyanate, 50 mM Tris-HCl (pH 7.5), 10 mM EDTA, 1.12 M 2-mercaptoethanol) at a concentration of 3×10^6 mammalian cells/ml lysis solution.

2. Lysis is taken to completion by solubilizing the tissue with a Polytron blender which also shears DNA present in the extract.
3. A chloroform–isoamyl alcohol (24:1) extraction with a volume equivalent to the extract volume is performed by vortexing sample for 15 sec and centrifugation at 2500 *g* for 10 min at room temperature.
4. The upper (aqueous) phase is transferred to a fresh tube containing 1.4 vol. 6 M LiCl. RNA is precipitated at 4°C for at least 15 h before centrifugation at 10 000 *g* for 30 min at room temperature.
5. The supernatant is removed by aspiration and the pellet is resuspended in half the volume of lysis buffer used previously (PK buffer: 50 mM Tris-HCl (pH 7.5), 5 mM EDTA, 0.5% (w/v) SDS) supplemented with 200 μg/ml proteinase K. Digestion of the proteins in the extract is achieved by incubation at 45°C for 30 min.
6. Sodium chloride is added to a final concentration of 0.3 M by adding 1/10 vol. 3 M NaCl.
7. Three phenol–chloroform–isoamyl alcohol extractions (step 3) remove the protein contaminants from the RNA, centrifuging each time at 2500 *g* for 10 min at room temperature.
8. The RNA in the recovered aqueous phase is precipitated with 2.5 vol. ethanol at −20°C for 2 h. RNA is pelleted by centrifugation at 10 000 *g* for 15 min at 4°C before being resuspended in a volume of solution appropriate for further procedures.

2.6 Fractionation of small RNA species using HPLC

The isolation of small RNA is not generally possible with the methods described previously. Many of the RNA molecules of interest (tRNA, 5S rRNAs and snRNAs) fall into this category and can be fractionated from total RNA using HPLC. HPLC is a method of quantitative separation for components in a sample solution utilizing a column of solid stationary phase material and a mobile liquid phase. The choice of the column and liquid phase depends entirely on the sample to be separated. The columns are normally packed with proprietary synthetic resins and the liquid phase solution (eluant) can be of constant concentration (isocratic) or can change in composition over time (gradient mode).

HPLC can separate nucleic acids by use of various characteristics: their polyanionic nature (anion-exchange); lipophilic nucleobases

LIVERPOOL JOHN MOORES UNIVERSITY
LEARNING SERVICES

which affect behavior on reversed-phase columns; or chain length. Small RNAs may be separated with reversed-phase or anion-exchange systems. Samples are collected in fractions and absorbance at 260 or 270 nm quantitates the RNA concentration in each fraction.

2.6.1 Fractionation using anion-exchange chromatography

In this type of chromatography, the species of RNA are separated on the basis of initially being bound to the support matrix which is positively charged and then progressively eluted by the increasing ionic strength of the eluant. Generally, an RNA species of high negative charge is hardest to elute and will elute from the column only when the eluant is of high ionic concentration (e.g. 1 M ammonium phosphate). Larger RNA species are therefore most tightly bound to the column matrix giving separations almost entirely based on size fractionation.

A system that has been used for isolation of low concentration tRNA species (and also applicable to snRNA) has been reported [7] and is described here (Protocol 2.6).

Protocol 2.6: Fractionation using anion-exchange chromatography

1. The solid phase consists of a prepacked column (5 μm particle size, 250 × 4.6 mm internal diameter W-porex 5 C-4, Phenomenex (Rancho Palos Verdes, CA).
2. The eluant used is a decreasing gradient (100–70%) of ammonium sulfate over 2 h at pH 7.0 or 4.5. Aminoacylated tRNAs can be separated at pH 4.5 but with lower resolution than that achieved for tRNAs at pH 7.0.
3. Absorbance at 270 nm or measurement of labeled tRNA is used to define concentration in the various fractions and to define the peaks that are associated with specific RNA. *Figure 2.3* illustrates the fractionation of $tRNA^{Leu}$ at pH 4.5.

2.6.2 Fractionation using reversed-phase chromatography

Separation of RNA species can be based on the lipophilic character of nucleic acid bases. RNA can be bound to the non-polar column matrix by hydrophobic interactions and is eluted by organic solvents. Separation is not strictly due to size alone as the base composition of the RNA has a large effect on the binding to the column. The

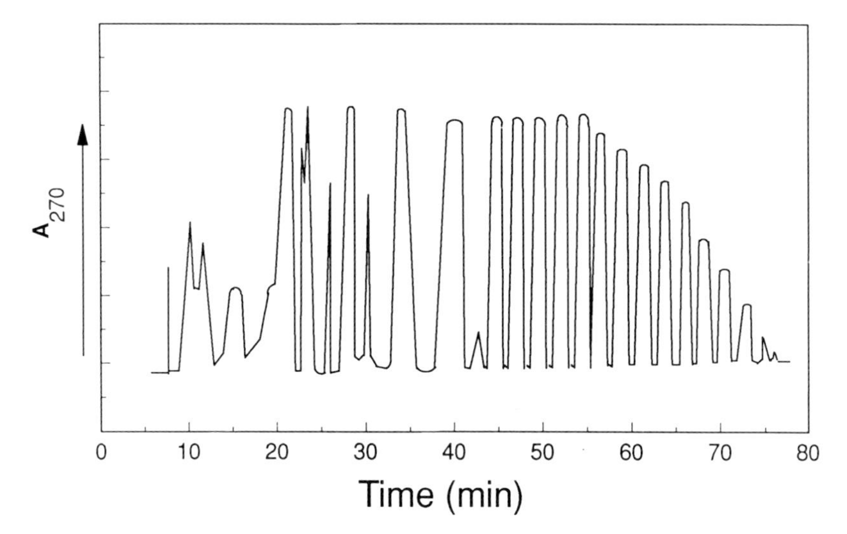

Figure 2.3: HPLC anion-exchange separation of RNA. Preparative anion-exchange HPLC of an oligo(rA) preparation on a PEI column. Oligo(rA) (5.6 mg) on a Baker PEI wide pore column (4.6 × 250 mm). Bruker LC21 B HPLC coupled to a Shimadzu SPD-6A spectrophotometer.

chromatographic behavior of the sample using reversed-phase chromatography is therefore harder to predict and is usually used to estimate purity rather than for identification procedures.

A system using a C_{18} column can be used for identifying and fractionating the products of RNase digestion of tRNA [8]. The eluant used is a gradient of 0–10% (v/v) methanol for 5 min, 10–25% (v/v) methanol for 20 min in the presence of 0.05 M ammonium phosphate, pH 4.5 at a flow rate of 4 ml/min. This separates digestion fragments yielding individual species in highly defined fractions (*Figure 2.4*).

2.7 Isolation of RNA from small numbers of cells

Isolation of RNA from a small population of cells has become possible by integration of standard techniques with affinity procedures. Many commercial kits are now available, some of which are listed in *Table 2.4*.

The methodologies of the kits vary and only one is described here. The GlassMAX RNA microisolation Spin cartridge system incorporates an extraction step based upon the guanidium thiocyanate method for the initial stages of extraction followed by selective binding of RNA to silica-based particles in the presence of high salt concentrations (Protocol 2.7).

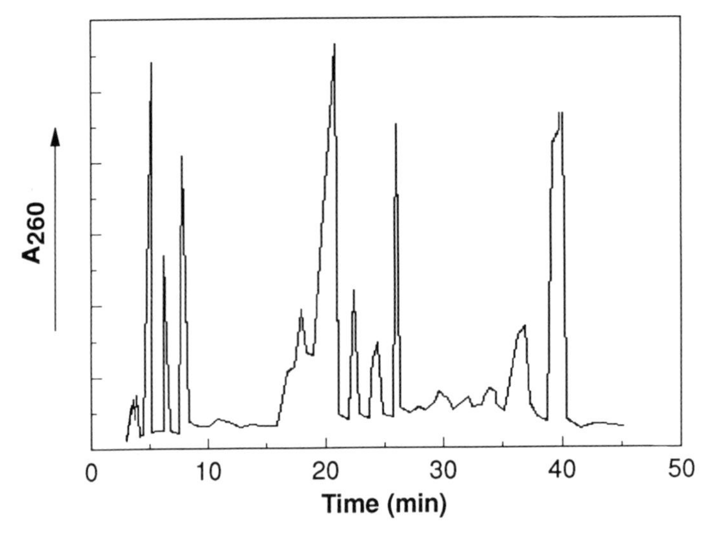

Figure 2.4: HPLC reversed-phase separation of RNA. Semi-preparative anion-exchange HPLC on a RP-NH_2 column of a nuclease digest of an aminoacyl-tRNA. Three A_{260} units of RNase T1 digest of Phe-tRNA on a LiChrosorb RP-NH_2 column. A DuPont 850 liquid chromatograph with a DuPont UV spectrophotometer was used to generate the separation.

Protocol 2.7: Isolation of RNA using the GlassMAX spin cartridge system

1. Approximately 5–20 mg of tissue can be homogenized in 400 µl 3.68 M guanidinium thiocyanate, 1.12 M 2-mercaptoethanol. Homogenization can be achieved by hand using a pestle that fits in a microcentrifuge or with a microprobe extension of a Polytron blender.
2. RNA is precipitated with 0.7 vol. ethanol at −20°C for 30 min and pelleted by centrifugation at 13 000 *g* for 5 min at 0°C.
3. The supernatant is removed by aspiration and the pellet dissolved in 450 µl 6 M sodium iodide (binding buffer).
4. Next 40 µl 3 M sodium acetate is added before the solution is loaded on to a GlassMAX spin cartridge which fits into a microcentrifuge tube.
5. Centrifugation at 13 000 *g* forces the sample through the column leaving RNA bound to the matrix material. This also removes some contaminants into the resultant throughflow.
6. The column is washed twice by addition of 0.5 ml ice-cold 80% (v/v) ethanol. Residual washing solution is removed by centrifugation at 13 000 *g* for 1 min.
7. Elution of the RNA is achieved by addition of 40 µl DEPC-treated water preheated to 65°C and centrifugation at 13 000 *g* for 20 sec, recovering the RNA into a sterile microcentrifuge tube.

Table 2.4: Microscale isolation kits for RNA

Kit name	Supplier	Comment
GlassMAX RNA microisolation Spin cartridge system	Gibco–BRL	Total RNA purified by affinity of RNA to silica-coated membrane (method given in the text)
Magnetic RNA isolation kit	Dynal	Oligo(dT)-bead based affinity for mRNA
RNAgents® Total RNA isolation system	Promega	Total RNA isolation kit
PolyATtract® system 1000	Promega	mRNA isolation using magnetic beads
USB Poly(A)RNA isolation kit	United States Biochemical (USB)	mRNA isolation kit using oligo(dT) cellulose columns
REX™	United States Biochemical (USB)	Total RNA extraction based on LiCl method

RNA isolated in these microprocedures often has a higher recovery relative to starting material mass than other procedures. RNA prepared in this manner is suitable for providing template material for reverse transcriptase–polymerase chain reaction procedures (RT–PCR) (Section 5.5.1) or for isolating RNA synthesized *in vitro* (Section 2.10).

2.8 Isolation of RNA using ultracentrifugation

Owing to the hazards inherent in working with phenol, a number of other methods have been investigated. Efforts have also been made to devise a simple method that will avoid the problem of DNA contamination. Both these objectives can be achieved by using centrifugation for isolating RNA. RNA species of defined size such as tRNA and snRNA can be isolated from total RNA preparations by rate-zonal density gradient centrifugation (Section 3.2) on some gradients or by preparative gel electrophoresis (Section 3.3).

In solutions containing high concentrations of cesium salts, RNA has a much higher density than DNA, protein or polysaccharide. Under the appropriate centrifugation conditions, RNA will pellet while the contaminants remain in the supernatant. One of the original methods involved lysis of cells with SDS followed by the addition of CsCl solution and ultracentrifugation [4], however, cesium dodecyl sulfate formed during this procedure is insoluble and must be removed before ultracentrifugation.

The most popular preparative centrifugation procedure for preparing total cellular RNA is detailed in Protocol 2.8.

Protocol 2.8: Isolation of RNA using ultracentrifugation

1. The tissue is homogenized in a solution of 4 M guanidinium thiocyanate, 30 mM sodium acetate, 0.14 M 2-mercaptoethanol [5]. Guanidinium thiocyanate is chaotropic and rapidly disrupts nucleoproteins into their component molecules and at the same time denatures all the proteins. 2-Mercaptoethanol reduces disulfide bonds of proteins, ensuring their complete denaturation.
2. The cell homogenate is layered over 1/2 vol. 5.7 M CsCl and centrifuged in a swing-out rotor at 250 000 *g* for 3 h at 22°C (remember that the use of these dense solutions will probably require the maximum speed of the rotor to be reduced. The DNA, polysaccharides and proteins remain above the density barrier while the RNA is pelleted giving a yield of about 90%.
3. The pelleted RNA is washed twice with 70% (v/v) ethanol, 0.3 M sodium acetate and then dissolved in sterile water previously treated with 0.1% (w/v) diethylpyrocarbonate (DEPC).

2.9 Fractionation of specific types of RNA using affinity methods

RNA populations isolated using one of the previous methods can be further fractionated to obtain specific types of species of RNA. Affinity isolation of RNA depends on the RNA possessing some feature that will distinguish it from both other macromolecules and, if possible, other types of RNA. The most frequent type of affinity isolation procedure is the purification of eukaryotic mRNA by virtue of the poly(A) tail which is a distinguishing feature of many types of mature mRNA. Large hnRNAs also contain 30-nucleotide long poly(U) segments [9]. Given the importance of isolating mRNA for all types of sequence and functional analyses, it is not surprising that considerable effort has been put into developing affinity methods for purifying mRNA.

An alternative approach for the affinity purification of RNA, applied to newly synthesized RNA, is to include a nucleotide analog as a specific marker before extraction. Typically, the RNA is synthesized *in vitro* in the presence of a nucleotide which has a chemical group that allows it to be distinguished from all the other types of RNA. Nucleotide analogs, used to differentiate and isolate the newly synthesized RNA, have included sulfhydryl [6] and biotin [10] groups which bind to columns with immobilized mercury and streptavidin (avidin), respectively.

2.9.1 Fractionation of mRNA using oligo(dT) cellulose

One of the first types of mRNA affinity column used were oligo(U) cellulose columns, but subsequently oligo(dT) columns became favored due to their greater stability; oligo(dT) cellulose is available commercially from several suppliers. In this procedure, total cellular RNA prepared as described in Sections 2.2 or 2.3 is loaded on to the column in a high salt load buffer (e.g. 1.0 M NaCl, 0.5% SDS, 1 mM EDTA, 10 mM Tris-HCl, pH 7.5). The amount of RNA to be loaded depends on its availability and the experimental needs. If the tissue is easily obtained then large amounts of total RNA can be prepared from which mRNA can be isolated. However, when tissue is not plentiful only small amounts of mRNA can be isolated. When large amounts of mRNA are required the quantity of oligo(dT) cellulose used in the experiment must also be increased. As a guide, 1 mg of total RNA requires 25 mg of oligo(dT) cellulose. The oligo(dT) cellulose can be used to create a column in two ways (Protocols 2.9 and 2.10).

Protocol 2.9: Batch adsorption of RNA oligo(dT) cellulose

1. Oligo(dT) cellulose may be mixed with the RNA solution obtained from the phenol–chloroform–isoamyl alcohol extraction (Sections 2.2 and 2.3). As a guide the amount of oligo(dT) cellulose added is approximately 1/20 of the original tissue weight.
2. The salt concentration of the RNA solution is then raised to 0.5 M by addition of 1/10 vol. 5 M NaCl. This favors hybridization of mRNA poly(A) tails to the oligo(dT) cellulose.
3. The solution containing the oligo(dT) cellulose is gently rocked at room temperature for 15 min to ensure that all of the mRNA hybridizes to the oligo(dT) cellulose.
4. The oligo(dT) cellulose is then collected by gentle centrifugation at 75 *g* for 1 min, without the brake.
5. The supernatant is removed by aspiration and the pellet is resuspended in 3 vol. loading buffer. The loading buffer (0.5 M NaCl in Tris-EDTA buffer) also has a high salt concentration which ensures the stability of the hybrids.
6. Centrifugation at 75 *g* for 1 min without the brake pellets the oligo(dT) cellulose leaving contaminants in the supernatant.
7. The remaining contaminants are removed by three further washes: resuspension in 4 vol. loading buffer and centrifugal collection as detailed above.
8. The oligo(dT) cellulose is then resuspended in 3 vol. loading buffer and loaded into an autoclaved polypropylene column. The oligo(dT) cellulose column bed forms by gravity flow as the loading buffer drips from the column.

9. Further washing of the formed column is achieved by addition of 10 bed vol. loading buffer, which is again allowed to drain under gravity. The effluent may be collected and passed back through the column to achieve a better recovery.
10. The poly(A)$^+$ mRNA is eluted by addition of 4 bed vol. elution buffer (Tris-EDTA (TE) buffer) preheated to 65°C. The combination of heat and the absence of salt leads to mRNA detaching from the oligo(dT) cellulose. The eluant is collected in a prechilled, siliconized glass centrifuge tube. The RNA eluted may still contain some rRNA which can be removed by passing the RNA back through the column, washing and eluting as before.
11. If the RNA is to be immediately precipitated then 1/10 vol. 3 M sodium acetate (pH 5.6) and 2 vol. ice-cold ethanol are added and left to precipitate at −20°C overnight. The precipitation overnight ensures a good recovery of the RNA.
12. The precipitated mRNA can be pelleted by centrifugation at 14 000 *g* at 4°C for 10 min and the pelleted RNA is then dissolved in 500 μl sterile water.
13. The RNA is transferred from the siliconized glass centrifuge tube to a microcentrifuge tube.
14. An equal volume of phenol–chloroform–isoamyl alcohol is added to the RNA and shaken to extract any remaining proteins. The phenol–chloroform–isoamyl alcohol is separated from the RNA by centrifugation at 10 000 *g* for 5 min at room temperature.
15. The recovered supernatant is then subject to a chloroform extraction with an equal volume of chloroform–isoamyl alcohol to remove any phenol carried over from the prior extraction.
16. The RNA is then precipitated by the addition of 1/10 vol. 3 M sodium acetate (pH 5.6) and 2 vol. ice-cold ethanol at −20°C for 2 h. Centrifugation at 10 000 *g* for 10 min at 0°C pellets the RNA which is then washed with 200 μl ice-cold 70% (v/v) ethanol to remove any salt present.
17. The RNA is dissolved in a volume of solution appropriate for further analysis.

Protocol 2.10: Loading total RNA on to oligo(dT) cellulose columns

Columns may be used to load high-purity RNA prepared by use of an isolation method such as those in Sections 2.2 and 2.3.

1. Columns are formed by addition of the required amount of oligo(dT) cellulose to 5 ml regeneration solution followed by loading into a sterile polypropylene column. As a guide the

amount of oligo(dT) cellulose added is approximately 1/20 of the original tissue weight.

2. Once the column has formed by gravity flow it is washed with sterile water until the effluent has a pH below 8.0.
3. The column is then equilibrated with 3 vol. loading buffer (0.5 M NaCl in Tris-EDTA buffer).
4. RNA dissolved in 0.5 M NaCl up to a maximum concentration of 5 mg/ml is loaded on to the column.
5. Three column volumes of loading buffer are used to wash contaminants from the column.
6. RNA can then be eluted by addition of 3 column vol. elution buffer (Tris-EDTA) preheated to 65°C.
7. If the RNA is to be immediately precipitated then it is collected upon elution into ice-cold tubes and precipitated with 1/10 vol. 3 M sodium acetate and 2 vol. ice-cold ethanol. Precipitation overnight at −20°C ensures a high recovery of RNA.
8. Centrifugation at 10–14 000 *g* pellets the RNA precipitate which can be washed with 500 μl aliquots of 70% ethanol to remove contaminating salt.
9. Further purification can be performed to remove contaminating rRNA by passing the eluant back through the column, washing and eluting as before.

Protocol 2.11: Regeneration of the oligo(dT) cellulose

1. The oligo(dT) cellulose can be regenerated for future use by washing it with a regeneration solution of 0.1 M NaOH, 5 mM EDTA. This is often desirable due to the cost of oligo(dT) cellulose. Oligo(dT) cellulose is resuspended in 10 bed vol. regeneration solution and transfered to a centrifuge tube.
2. The oligo(dT) cellulose is pelleted by centrifugation at 75 *g* for 1 min, without the brake.
3. Three further washes with 10 vol. regeneration solution and collection by centrifugation removes contaminants.
4. Two washes with 3 vol. load buffer with centrifugal collection removes NaOH from the oligo(dT) cellulose.
5. Resuspension in 1 vol. load buffer and storage at 4°C leaves the oligo(dT) cellulose ready for future use.

2.9.2 Fractionation of RNA using magnetic beads

Columns are prone to losses of RNA on the supporting matrix; they also tend to become clogged and so may flow very slowly. Instead of using a column, methods have been devised using magnetic beads.

There are a number of variations upon the method which involve binding of beads coated either with oligo(dT) for isolating total mRNA or with a defined complementary sequence for isolating a specific RNA. The oligonucleotide probes can be bound directly to the beads or via biotin to streptavidin-coated beads. Streptavidin can bind more than one biotin per molecule giving a better sequestration of mRNA.

The RNA sample can be prepurified total cellular RNA (Sections 2.2 and 2.3) or a total cell lysate. For example, cells can be lysed by the addition of guanidinium thiocyanate homogenization solution which contains biotinylated oligo(dT). The yield of mRNA using this technique tends to be higher and the length of time required is less because the hybridization takes place in liquid phase rather than solid phase required by the oligo(dT) column methodology.

Biotinylation of oligonucleotides. Many affinity techniques involve the use of biotinylated oligonucleotides. Biotin can be attached to RNA chemically or enzymatically. Chemical methods use either incubation of oligonucleotides carrying amino-group linkers with solid sulfo-*N*-hydroxysuccinimide long-chain biotin or photobiotinylation where the linking step is achieved by photochemical reactions.

Protocol 2.12: Chemical biotinylation of RNA

1. The chemical incubation procedure uses 35 µg of an oligonucleotide and 0.3 mg of solid sulfo-*N*-hydroxysuccinimide long-chain biotin $(CH_2)_9$ with 20 µl carbonate buffer (50 mM $NaHCO_3$/100 mM Na_2CO_3, pH 9.0) [12]. The mixture is incubated at room temperature for 4 h and dried *in vacuo*.
2. Purification of the biotinylated oligonucleotide is achieved by denaturing polyacrylamide gel electrophoresis (PAGE) using a 20% (w/v) polyacrylamide/8 M urea gel (Section 3.3.3).

Chemical biotinylation of oligonucleotides can also be achieved by use of an oligonucleotide synthesizer which can incorporate a biotinylated nucleotide during synthesis. This is a common method of generating tagged oligonucleotides used particularly for polymerase chain reaction (PCR) systems.

Enzymatic biotinylation can be achieved by incorporation of biotinylated nucleotides into nucleic acid chains. This is mediated by the large fragment of DNA polymerase I (Klenow fragment) which incorporates a biotinylated dUTP (Clontech) on to the end of a DNA fragment released from a vector. Labeling by incorporation is more efficient than chemical biotinylation.

Protocol 2.13: Enzymatic biotinylation of RNA

1. The fragment must be released from the vector with a restriction enzyme that leaves an overhang which includes a thymidine residue (e.g. *Bam*HI, *Eco*RI, *Hin*dIII, *Kpn*I, *Pst*I, *Pvu*I, *Sst*I, *Xba*I and *Xho*I).
2. Klenow and nucleotide mix (bio21UTP, dATP, dCTP and dGTP) is added to the restricted oligonucleotide and incubated at 37°C for 10 min. This allows Klenow to 'end fill' the fragment incorporating the biotinylated nucleotide added.
3. Unincorporated bio21UTP can be removed from the RNA product by use of a Sephadex G50 spin-column.

Use of oligo(dT)-coated magnetic beads for mRNA isolation. In this method the mRNA binds to the oligo(dT) attached to magnetic beads. Oligonucleotides can be attached to the beads in various ways. One method achieves this by mixing 5 mg of amino beads (Dynabeads L255 or R469) with 600 pmol oligonucleotide in 1 ml of 0.1 M imidazole buffer (pH 7.0), 0.1 M carbodiimide (EDC) [13]. The mixture is incubated at 50°C for 3 h with gentle shaking.

Protocol 2.14: mRNA isolation using oligo(dT)-coated magnetic beads

1. The tissue is homogenized in a guanidinium thiocyanate buffer which is adjusted by dilution (62.5 ml water/100 ml 4 M guanidinium thiocyanate extract) to 2.5 M guanidinium thiocyanate, 0.2 M Tris-HCl (pH 7.5), 0.04 M EDTA, 0.5% (w/v) Sarkosyl and 10% (w/v) dextran sulfate-5000 [11].
2. A 0.5 ml aliquot of oligo(dT)-coated beads (1 mg of beads carrying 62.5 pmol oligo(dT), a 5–6 molar excess) in bead prehybridization buffer (0.1 M Tris-HCl (pH 7.5), 0.01 M EDTA, 4% (w/v) fraction V bovine serum albumin (BSA), 0.5% (w/v) sodium lauroyl sarcosine, 0.05% (w/v) bronopol.)) is mixed with 0.25 ml crude extract by vortexing for 10 sec. This lowers the guanidinium thiocyanate concentration to 0.83 M.
3. Hybridization of the mRNA to the oligo(dT) occurs for 5 min at 37°C.
4. Beads are harvested by use of a neodymium–iron–boron permanent magnet, discarding the clear supernatant and resuspending the hybrid–bead complexes in a solution suitable for further procedures.

Isolation of RNA using biotinylated oligonucleotides. In this method the RNA is hybridized to the probe and this complex is subsequently attached to streptavidin-coated magnetic beads. An example of this type of procedure is shown in *Figure 2.5*. Commercial kits are available which contain biotinylated oligo(dT) that can hybridize to mRNA in total cell RNA extracts. In the isolation procedure the poly(A) tails of mRNA bind to the biotinylated oligo(dT) and the hybrid, in turn, is bound to beads coated with streptavidin. The magnetic beads can be collected using a magnet, washed and the RNA eluted using a low-salt buffer.

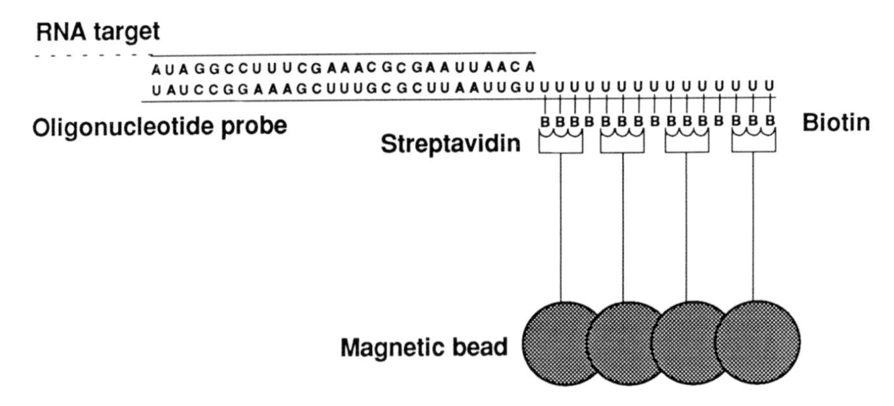

Figure 2.5: Hybrid capture of biotinylated probe–target complexes using streptavidin-coated magnetic beads.

Hybridization of RNA species to biotinylated oligonucleotide. Biotinylated oligonucleotides complementary to an RNA type or species can be used during isolation procedures or after total RNA isolation procedures. Oligo(dT) nucleotides or oligonucleotides complementary to a specific RNA sequence can be biotinylated (see earlier) and can be used to hybridize to RNA. The amounts of oligonucleotide used for hybridization to mRNA varies with respect to RNA type. Capped mRNA (Section 1.6.1) is hybridized to 0.4 μg biotinylated oligonucleotide per 0.25 μg capped mRNA. Radiolabeled [^{32}P]mRNA is hybridized to 0.1 μg biotinylated oligonucleotide per 80 ng [^{32}P]mRNA [14]. Hybridization in 50 μl hybridization buffer (1 M NaCl, 50 mM piperazine-*N*,*N*′-bis(ethanesulfonic acid) (PIPES; pH 7.0), and 2 mM EDTA) is carried out at 85°C for 2 min and then at 55°C for 1 h.

Hybrid capture on streptavidin-coated magnetic beads. The biotinylated hybrid can be captured by magnetic beads coated with streptavidin. Streptavidin-coated beads are commercially available (Dynabeads M-280 streptavidin, Dynal) where the streptavidin is covalently linked to the beads.

Protocol 2.15: Hybrid capture on streptavidin-coated magnetic beads

1. Dynabeads are washed before use in 1 M NaCl and TE buffer.
2. 20 μg biotinylated DNA fragment-captured RNA hybrid are mixed with 25 mg prewashed Dynabeads in a volume of 150 μl TE buffer.
3. Hybridization occurs at room temperature for 30 min with constant mixing on a Coulter/end-over-end mixer.
4. The bead–biotinylated hybrid can be collected by use of a neodymium–iron–boron permanent magnet, discarding the supernatant and resuspension of the hybrid–bead complexes in a solution suitable for further procedures (e.g. water for PCR analysis).

Protocol 2.16: Hybrid captured using streptavidin agarose

1. The streptavidin agarose is prepared by three prewashes of 60 μl streptavidin agarose suspension (Sigma Chemical Co.) in 0.5 ml wash buffer (10 mM Tris-HCl, pH 8.0, 0.25 M NaCl).
2. A further 100 μl wash buffer supplemented with 5 μg calf liver tRNA (Boehringer Mannheim) is mixed on a rotary shaker for 15 min at 300 r.p.m to block non-specific hybridization.
3. The hybridization mixture (biotinylated oligonucleotide:mRNA hybrid in hybridization buffer) is added to the streptavidin agarose along with 50 μl sterile double-distilled water.
4. Binding of the DNA:RNA hybrid to streptavidin–agarose occurs whilst shaking in a rotary incubator at 300 r.p.m for 90 min.
5. Agarose is separated from the suspension by centrifugation in a Spin-X tube (Costar) for 3 min in a microcentrifuge. The agarose is washed twice with 150 μl sterile distilled water to remove contaminants.
6. The mRNA is recovered from the agarose by addition of 200 μl elution buffer (10 mM Tris-HCl (pH 7.8), 30% deionized formamide). Incubation at 60°C for 10 min precedes immediate quick freezing in a dry ice–ethanol bath.
7. The sample is slowly thawed in ice before centrifugation in a microcentrifuge Spin-X tube for 3 min. mRNA is eluted in the flow-through fraction and can be precipitated with 1/10 vol. 3 M sodium acetate (pH 5.2), 2.5 vol. ethanol at −20°C for 20 min.
8. Centrifugation for 3 min in a microcentrifuge pellets the RNA and a wash with ice-cold 70% ethanol removes excess salts. The pellet is dried *in vacuo* and dissolved in a suitable amount of sterile water (Section 2.3).

2.9.3 Isolation of newly synthesized RNA by affinity methods

There are a number of *in vitro* transcriptional systems used for analyzing a range of factors involved in the control of gene expression. In some cases, for example when using isolated nuclei as a transcription system, there is already a large amount of pre-existing RNA and so the RNA that is synthesized *in vitro* will only represent a tiny percentage of the total RNA. Hence, in order to characterize the newly synthesized RNA it must be separated from the pre-existing RNA. All the methods devised for the isolation of newly synthesized RNA depend on the incorporation of modified nucleotides which can be used as a means of distinguishing and isolating the RNA that has been synthesized *in vitro*.

Incorporation of nucleotide analogs into RNA. One approach is to produce labeled transcripts *in vitro* using mercurated nucleotides and subsequently select these using sulfhydryl-Sepharose affinity chromatography [15]. Alternatively, *in vitro* labeling of polynucleotides with phosphothioate analogs of nucleoside triphosphates has been used in conjunction with organomercury affinity chromatography to study RNA chain initiation, capping and enzyme–polynucleotide interactions [16]. 4-Thiouridine (4-TU) and 6-thioguanosine (6-TG) triphosphates (*Figure 2.6*) are normally used for this purpose. The newly synthesized RNA is usually also radiolabeled by the addition of 10–40 kBq/ml RNA precursor simultaneously with 4-TU or 6-TG as a marker for uptake and incorporation into newly synthesized RNA and used to monitor RNA recovery during subsequent isolation and enrichment steps.

Protocol 2.17: Isolation of sulfhydryl-labeled RNA by affinity chromatography

1. After incubation the *in vitro* synthesized RNA is isolated (Sections 2.2 and 2.3) and redissolved in 50 mM sodium acetate, pH 5.5, 0.1% SDS, 0.15 M NaCl, and 4 mM EDTA.
2. The samples are heated to 65°C for 5 min to denature secondary structure, and cooled rapidly on ice to minimize the reformation of helical structures.
3. The denatured RNA is then batch-adsorbed to *p*-hydroxymercuribenzoate agarose (e.g. Affi-Gel 501, Bio-Rad) at 25–400 μg RNA/ml.
4. The non-specifically bound RNA is eluted with TE buffer containing 0.5 M NaCl and is discarded.
5. The thio-substituted RNA can be eluted with buffer containing 10 mM 2-mercaptoethanol.
6. Purification of newly synthesized RNA can be quantitated by determining the specific activity of ^{3}H-labeled RNA in c.p.m./μg RNA.

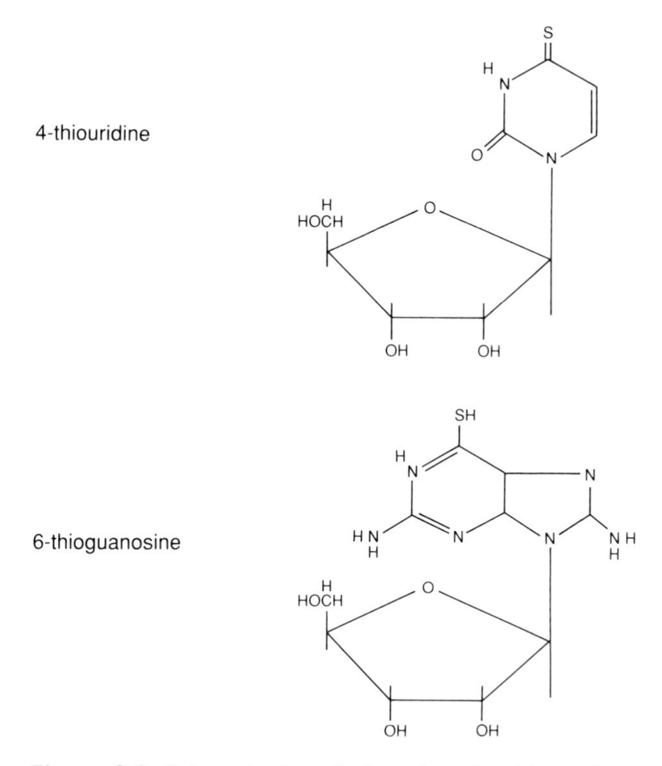

Figure 2.6: Selected thio-substituted nucleoside analogs.

Incorporation of biotinylated oligonucleotides into RNA. Another technique for the isolation of a specific RNA synthesized *in vitro* uses small, reversibly biotinylated RNAs for affinity chromatography [14]. An anchor DNA probe with sequences complementary to the RNA to be isolated is synthesized *in vitro*. The probe is conjugated molecules containing biotin at carbon-5 of a uridine residue via a linker containing a disulfide bond. The biotinylated anchor DNA is attached to the solid support via succinylavidin molecules each of which is capable of binding four biotin molecules. When the total RNA mixture is transferred into the column, the RNA of interest is hybridized to the anchor DNA via its complementary sequence. RNA that has not hybridized may be eluted by washing the column with buffer. The specifically bound RNA can then be eluted with buffer containing dithiothreitol (DTT) which reduces the disulfide bonds linking biotin to the anchor DNA.

2.10 Synthesis of RNA *in vitro*

RNA isolation procedures are complicated and often expensive. In order to bypass the need for isolation each time that RNA is required for RNA functional studies, RNA can easily be synthesized *in vitro*. RNA synthesis *in vitro* requires a DNA template with sequence complementarity to the RNA needed. This can be obtained either by screening genomic or cDNA libraries or by use of reverse transcriptase–PCR (RT–PCR) procedures (Section 5.5.1).

The DNA template must be flanked by transcription start sites (*Figure 2.7*) which are incorporated into many cloning vectors. Many plasmid vectors contain transcription sites (*Table 2.5*) which can be used to generate RNA molecules by addition of a specific RNA polymerase. Sequences generated by RT–PCR can have transcription sites tailored into the primers allowing transcription of the amplified sequences.

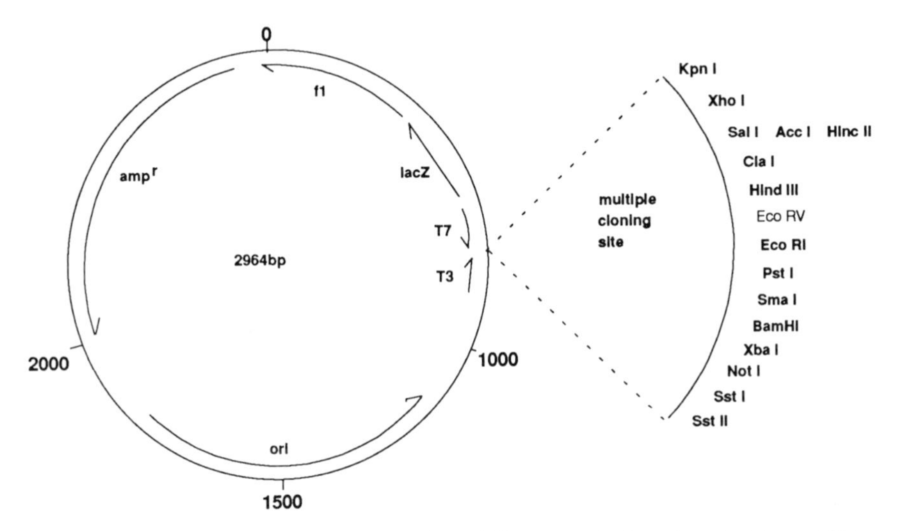

Figure 2.7: pBluescript (SK) vector. T3 and T7 transcription sites facilitate RNA synthesis of fragments inserted into the multiple cloning site.

Table 2.5: Selected vectors used for generation of RNA transcripts

Vector	Plasmid/phage	RNA polymerase sites	Comment
pGEM	Plasmid	SP6 and T7	
pBluescript (pBS)	Plasmid	T3 and T7	Major series of plasmid vectors used to derive RNA transcripts
Lambda Zap	Phage	T3 and T7	Insert sizes of up to 10.8 kb. Harbors a copy of pBS plasmid either side of the polylinker which allows excision of a subsection of the insert as a pBS vector containing desired inserts

The principle of the procedure is to add a RNA polymerase to the *in vitro* system and generate a RNA molecule. The binding of the RNA polymerases to their DNA-binding sites is highly specific such that *in vitro* systems produce the RNA species required without any contaminating transcripts. RNA transcripts are used in functional studies and for generating single-stranded radiolabeled-RNA probes (Section 3.5.1). Capped RNA can also be synthesized when using certain commercial kits (e.g. Ampliscribe SP6, T3 and T7, Cambio) by substitution of a portion of the GTP with the methylated or unmethylated forms of the cap analog ($m^7G[5']ppp[5']G$ or g[5′]ppp[5′]G, respectively).

Numerous vectors are used for transcription *in vitro*. A selection of vectors commonly used in molecular biology laboratories is given in *Table 2.5*. RNA transcripts must be generated from the insert only and not the plasmid DNA. Therefore, before transcription is carried out the vector must be linearized using restriction enzymes. The purity of the DNA has little effect on transcription making the use of crude minipreparations ('minipreps') possible which have been phenol–chloroform extracted to remove the protein fraction. The plasmid is linearized by use of restriction enzymes that cut at unique sites after the insert. The restriction enzymes favored for linearization generate 5′ protruding termini or blunt-ended DNA. Restriction generating 3′ protruding termini must be 'blunted' using the Klenow fragment of DNA polymerase I or T4 DNA polymerase. The digestion should be checked on ethidium bromide gels (Section 3.3.4) to ensure completion.

Protocol 2.18: *In vitro* transcription

1. Transcription mixture:
   a. transcription buffer 40 mM Tris-HCl (pH 7.5)
      5 mM NaCl
      6 mM $MgCl_2$
      2 mM spermidine HCl
      10 mM DTT
      100 μg/ml BSA (Fraction V; Sigma)
   b. 500 μM of each rCTP, rGTP, rUTP
   c. 20 nM linearized plasmid DNA
   d. 2–20 μM [α-^{32}P]rATP (specific activity of 14.8–111 TBq (400–3000 Ci)/mmol)
   e. 1 U/μl placental RNase inhibitor (e.g. RNasin™, Promega)
   f. 500 U/ml bacteriophage DNA-dependent RNA polymerase
2. Transcription proceeds for 1–2 h at 37°C (T3 and T7 DNA-dependent RNA polymerases) or 40°C (SP6 DNA-dependent RNA polymerase).

3. The DNA template is then removed by incubation with 1 μl RNase-free pancreatic DNase I (1 mg/ml) at 37°C for 15 min.
4. DNase I is removed by phenol–chloroform–isoamyl alcohol extraction recovering the aqueous phase to a fresh tube.
5. RNA is precipitated by addition of 1/5 vol. 5 M ammonium acetate and 2.5 vol. ethanol at −20°C for 30 min.
6. RNA is pelleted by centrifugation at 12 000 *g* for 10 min at 4°C and can be redissolved in water and used in further procedures.

References

1. **Qureshi, S.A. and Jacobs, H.T.** (1993) *Nucl. Acids Res.*, **21**, 811.
2. **Berthier, F., Renaud, M., Alziari, S. and Durand, R.** (1986) *Nucl. Acids Res.*, **14**, 4519.
3. **Kirby, K.S.** (1968) *Methods in Enzymology,* XIIB (L. Grossman and K. Moldave, eds) Academic Press. p. 87.
4. **Baulcome, D.C. and Buffard, D.** (1983) *Planta,* **157**, 493.
5. **Wilkinson, M.** (1991) in *Essential Molecular Biology: a Practical Approach* (T.A. Brown, ed.). IRL Press, Oxford. p. 69.
6. **Chomczynski, P. and Saachi, N.** (1987) *Anal. Biochem.*, **162**, 156.
7. **Dudock, B.S.** (1987) in *Molecular Biology of RNA: New Perspectives* (M. Inouye and B. Dudock, eds). Academic Press, New York. p. 321.
8. **Pingoud, A., Fliess, A. and Pingoud, V.** (1989) in *HPLC of Macromolecules: a Practical Approach* (R.W.A. Oliver, ed.). IRL Press, Oxford. p. 183.
9. **Molloy, G.R., Jelinek, W., Salditt, M. and Darnell, J.E.** (1974) *Cell,* **1**, 43.
10. **Lamond, A.I. and Sproat, B.S.** (1993) in *RNA Processing: a Practical Approach* (S.J. Higgins and B.D. Hames, eds). IRL Press, Oxford. p. 103.
11. **Morrissey, D.V., Lombardo, M., Eldredge, J.K., Kearney, K.R. and Patrick Groody, E.** (1989) *Anal. Biochem.*, **181**, 345.
12. **Kwoh, D. Y., Davis, G.R., Whitfield, K.M., Chappelle, H. L., DiMichele, L.J. and Gingeras, T.R.** (1989) *Proc. Natl Acad. Sci.* USA, **86**, 1173.
13. **Lund, V., Schmid, R., Rickwood, D. and Hornes, E.** (1988) *Nucl. Acids Res.*, **16**, 10861.
14. **Soh, J. and Pestka, S.** (1993) *Meth. Enzymol.,* **216**, 186.
15. **Feist, P.L. and Danna, K.J.** (1981) *Biochemistry,* **20**, 4243.
16. **Gilmour, R.S.** (1984) in *Transcription and Translation: a Practical Approach* (B.D. Hames and S.J. Higgins, eds). IRL Press, Oxford. p. 131.
17. **Melvin, W.T. and Keir, H.M.** (1979) *Anal. Biochem.,* **92**, 324.

3 Characterization of RNA size

3.1 Basic principles

RNA is a linear molecule usually possessing few branch points but with often high levels of secondary and tertiary structure (e.g. tRNA molecules). RNA molecules that do possess branch points will affect some methods of size determination, producing anomalous results. In addition, RNA molecules are also prone to aggregation, and this can complicate size determination.

Determinations of RNA size can be based on the migration of molecules in a centrifugal or electrical force field. These determinations are not definitive measurements of RNA size but are often quicker than a high-precision method afforded by sequence analysis. Sequencing methods will be only briefly discussed in this chapter as this methodology is fully described in Section 4.5.

3.2 Determination of the size of RNA using centrifugation

The use of centrifugation for determination of RNA size is of historical interest as it has now been largely superseded by electrophoretic procedures which give much better resolution. Many of the quicker centrifugal procedures for size assessment are still used during preparative centrifugal procedures. Preparative centrifugation also has higher RNA recovery rates than electrophoretic methods and thus it is important to understand the basis of centrifugal separation.

3.2.1 Sedimentation coefficients

The size of an RNA molecule is usually expressed in terms of its sedimentation coefficient which tends to be characteristic for some types of RNA. For convenience, sedimentation coefficients are expressed as Svedberg (S) units, where 1S is equal to 10^{-13} sec. This gives values that are easy to work with and, coincidentally, a particle of 1S is about the smallest that can be pelleted in an ultracentrifuge [1]. *Table 3.1* gives a list of s-values for a number of common molecules; the majority of RNA molecules in the cell have s-values between 4S and 30S. Note that the relationship between s-value and molecular weight is not linear, moreover the observed s-value depends on the conformation of the RNA molecules which are particularly affected by the ionic conditions. For example, in the presence of divalent cations RNA molecules become more compact and the observed s-value increases significantly. Hence, it is important to use standard ionic conditions when measuring the s-values of RNA.

Table 3.1: Sedimentation coefficients of selected RNAs

Type of RNA	s-value	Mol. wt (kDa)	Nucleotides
tRNA	4S	30	85
5S rRNA	5S	40	120
16S rRNA	16S	530	1776
23S rRNA	23S	1070	3566
18S rRNA	18S	710	2366
28S rRNA	28S	1900	6333
α-globin mRNA (mouse)	9S	220	696
β-globin mRNA (mouse)	9S	220	783

3.2.2 Analytical ultracentrifugation

Centrifugation, in either the analytical ultracentrifuge or later by rate-zonal centrifugation on sucrose gradients in preparative centrifuges was one of the first methods developed for determining the size of RNA. In the case of the analytical ultracentrifuge, the RNA is dissolved in a dilute buffer or salt solution and sedimented in a cell with two transparent windows which makes it possible to observe the formation of a moving boundary in the cell as the RNA sediments. Analysis of the rate of boundary movement as well as its shape can be used to deduce the size of the RNA. The detailed calculations used to determine sizes are complex but are based around the theory of sedimentation coefficient calculated by observing sample migration rates over time [1]. Further details may be found in standard textbooks that describe the methodology of analytical centrifugation [2].

The complexity, expense and limited applications of analytical ultracentrifuges led to numerous studies using swinging-bucket rotors in preparative ultracentrifuges. Here the separation of molecules is based on the use of a gradient. The gradient usually used is sucrose as the relationships between the density, temperature, concentration and viscosity of sucrose solutions are well known. Glycerol is also used as a gradient substrate but mostly for protein analysis.

3.2.3 Sucrose gradients

Two major features are noteworthy in measuring the sedimentation of RNA: firstly there is a limit to the resolution obtainable and secondly RNA molecules tend to aggregate in solution leading to anomalously high s-values. This latter phenomenon was the cause of much argument during the search for nuclear precursors of mRNA. The problem of RNA aggregation can be overcome by preparing denaturing sucrose gradients containing either 85% formamide or 99% dimethyl sulfoxide (DMSO) [3]; the former is less noxious and so is thus recommended. RNA can also be pretreated with formaldehyde [4], which prevents the amino groups forming hydrogen bonds and thus prevents secondary structure formation. One of the drawbacks is that sucrose is less soluble in these solvents and thus gradients must be shallower, making them less stable. The other feature that appears when RNA is centrifuged in denaturing gradients is that nicks in the molecule, hidden by the hydrogen-bonded secondary structure, are exposed as the RNA adopts a random coil conformation.

An example of the effect of a denaturing environment on gradient profiles is shown in *Figure 3.1*. Usually the ratio of 28S rRNA to 18S rRNA peaks should be 2:1, reflecting the fact that there are equal numbers of molecules but that the 28S rRNA is about twice as big as the 18S rRNA (*Table 3.1*). This is the case in non-denaturing

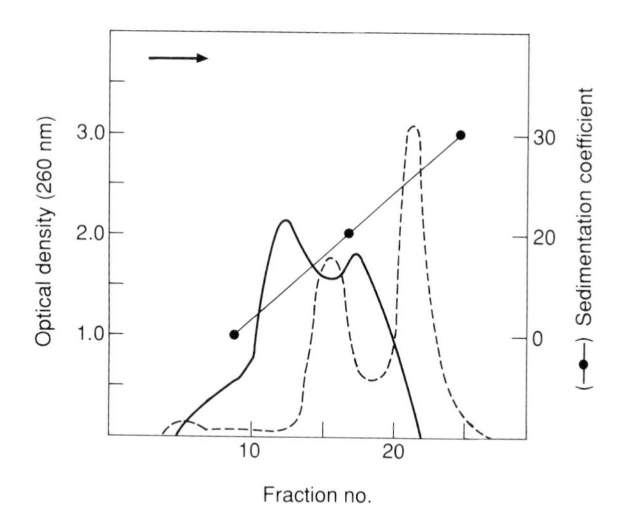

Figure 3.1: Sedimentation of ribosomal RNA in non-denaturing and denaturing sucrose gradients. Ribosomal RNA was isolated and centrifuged either on a non-denaturing aqueous sucrose gradient (- - - - -) or a denaturing formamide sucrose gradient (——). Note the loss of faster sedimenting RNA in the latter. Adapted from Ref. 3.

gradients but when the same RNA is centrifuged in formamide–sucrose gradients there is a marked change in the ratio of the two peaks (see *Figure 3.3*), indicating that a significant proportion of the RNA molecules are nicked.

Sedimentation coefficients can be determined by using either isokinetic sucrose gradients with markers of known s-values or one of the computer programs that are now widely available. *Table 3.2* lists some RNA markers that are commercially available.

Table 3.2: Commercially available RNA markers

Commercial description	Supplier	Comments
rRNA, 23S and 16S (*E.coli* R13)	Pharmacia	rRNA markers remain intact under
rRNA, 28S and 18S (calf liver)	Pharmacia	denaturing conditions; also used
rRNA, 25S and 18S (*S. cerevisiae*)	Pharmacia	as gel electrophoresis markers

Isokinetic gradients. Isokinetic gradients are gradients in which particles move at a constant speed, where the increasing centrifugal force on the particles, as they move away from the center of rotation, is balanced by the increasing viscous drag of the sucrose gradient. Fortuitously, linear gradients of 5–20% (w/w) sucrose at 5°C are essentially isokinetic. Usually, the sample is loaded on to the gradient together with marker RNA. Usually rRNA or tRNA make good markers, the choice of which depends on the likely size of the RNA molecule under investigation. Knowing the s-value of the marker (S_m), the distances moved by the marker (r_m) and the sample (r_s) allows the s-value of the sample (S_s) to be calculated from the equation:

$$S_s = (S_m \times r_s)/r_m$$

For maximum accuracy it is desirable to use two marker molecules, one larger than the unknown and the other smaller and then taking the average of the two calculations. The alternative is to use one of the computer programs which can calculate the s-value of RNA molecules irrespective of the profile and concentration range of the gradient. However, most computer programs only work with non-denaturing sucrose gradients.

RNA is layered as a narrow zone on the top of a linear gradient, usually of sucrose, originally with gradients of 5–20%(w/v) sucrose. It is important to avoid overloading the gradients otherwise the RNA molecules will sediment faster than expected. As a general guide, it is possible to load 100–200 μg of RNA on the top of a 12 ml 5–20% (w/v) sucrose gradient. In order to avoid overloading and to make subsequent analysis of the gradient simple, it is helpful to use radiolabeled markers since then little material is required.

The methods for analyzing sucrose gradients have been well established for many years. Gradients should be fractionated into at least 20 aliquots, preferably by upward displacement [2]. Upward displacement involves introducing a dense solution into the bottom of a gradient, pushing the banded samples higher, allowing the fraction collector to collect sequential samples as they are forced into the collection tube (*Figure 3.2*). The samples must then be analyzed. RNA absorbs light strongly at 260 nm and thus UV absorption is a good analytical method. Unfortunately, formamide and dimethyl sulfoxide also absorb in the UV and consequently gradients of these solvents cannot be analyzed for RNA concentration by measuring the optical density of fractions at 260 nm. If the RNA samples are radioactive then it is possible to measure the radioactivity of each of the fractions. Another measurement is possible if RNA of at least partially known sequence is being fractionated and a complementary labeled probe is available as the position of the RNA in the gradient can be found by hybridization methods (e.g. dot blots — Section 4.3.4).

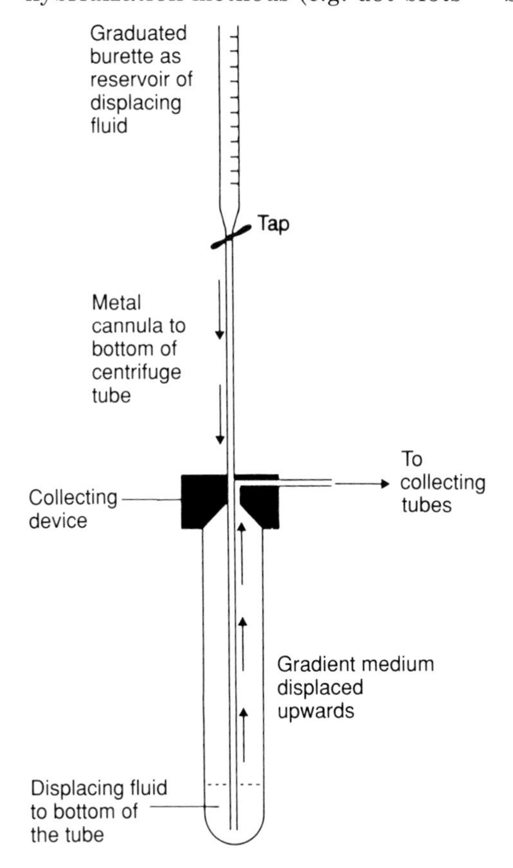

Figure 3.2: Unloading sucrose gradients by upward displacement. Reproduced from Ford and Graham (1991) *An Introduction to Centrifugation*, p. 48, BIOS Scientific Publishers Ltd.

In the final analysis, the degree of resolution that can be obtained using sucrose gradients is significantly less than can be obtained using electrophoretic procedures (*Figure 3.3*). Additionally electrophoretic analyses are simple, convenient and can be carried out with the minimum of effort and equipment. Hence, unless an environment that is incompatible with electrophoresis is to be used (e.g. high ionic strength), the usual choice would be to use electrophoresis for determining the sizes of RNA.

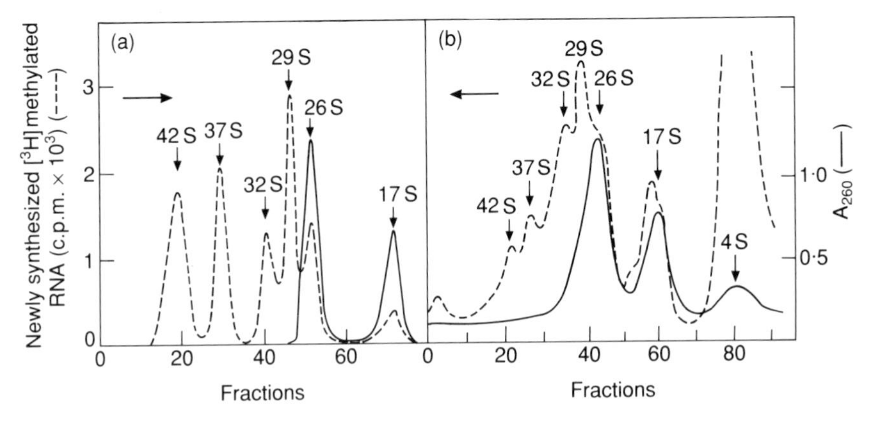

Figure 3.3: Comparison of resolution of gel electrophoresis and centrifugation. Separation of yeast total RNA by (a) polyacrylamide gel electrophoresis and (b) a rate-zonal sucrose gradient.

3.3 Determination of the size of RNA using gel electrophoresis

In the late 1960s Loening and his co-workers developed high-resolution gel electrophoretic methods for separating RNA which provide much higher resolution than can ever be achieved by centrifugation (*Figure 3.3*). Hence it is not surprising that gel electrophoresis has become the method of choice for determining the size of RNA. For this reason a wide range of analytical and microtechniques, including capillary gel electrophoresis (see Section 5.6), have been devised around this methodology that allows extensive analysis of quantities of RNA ranging from nanograms to milligrams [5].

3.3.1 Gel matrices

Polyacrylamide gels provide good separations and are convenient to use for almost all except the largest RNAs. Gel recipes suitable for separating RNA and the composition of the running buffers of choice are given in *Table 3.3*. The actual choice of gel concentration depends on the size of the RNA. For the largest types of RNA (>28S) the

Table 3.3: Recipes of polyacrylamide gels for separating RNA

	Final polyacrylamide concentration (%)					
	2.0	2.5	3.0	4.0	5.0	10.0
Acrylamide stock soln (ml)[a]	4.0	5.0	6.25	8.0	10.0	20.0
5 × TBE buffer (ml) (see Table 3.4)	6.0	6.0	6.0	6.0	6.0	6.0
Water (ml)	20.0	19.0	17.75	16.0	14.0	4.0
Final volume	30.0	30.0	30.0	30.0	30.0	30.0

[a]Acrylamide stock—30 g acrylamide, 1.5 g bisacrylamide made up to 200 ml with distilled water.

acrylamide concentration required is so low that the gels become fragile and difficult to handle. In these cases gels can be strengthened by the addition of 0.5% (w/v) of agarose and are commonly referred to as composite gels [6].

Polyacrylamide gel matrices are formed by polymerizing monomers of acrylamide with monomers of a cross-linking agent. The most commonly used cross-linking agent is *N,N′*-methylene-bis-acrylamide referred to as 'bis' for short. A three-dimensional matrix is formed where the concentration of acrylamide determines the average length of the polymer chains and the bis concentration defines the extent of cross-linking. The ratio of these monomers defines the eventual pore size, density, elasticity and mechanical strength of the gel. Other cross-linking agents are used for special purposes usually when gel segments are to be solubilized (RNA recovery). Agents such as *N,N′*-bis-acrylyl-cystamine (BAC) or diallyltartardiamide (DADT) allow gel solubilization in 2-mercaptoethanol and 2% periodic acid, respectively [5].

Polymerization of polyacrylamide gels is normally initiated by ammonium persulfate (APS) and the reaction is accelerated or catalyzed by *N,N,N′,N′*-tetramethylethylenediamine (TEMED). This is the favored initiation process as it gives reproducible homogeneous pore sizes throughout the gel. Another polymerization method utilizes riboflavin which triggers polymerization when the gel is exposed to UV radiation. This allows polymerization to be initiated after the gel is poured thus avoiding any possibility of premature polymerization. Polymerization is inhibited by dissolved oxygen in the gel mixture making it necessary to degas the gel mix before polymerization initiation. Degassing is routinely performed with water-suction pumps but rotary vacuum pumps can also be used. The concentration of the gel matrix need not be homogeneous, indeed gradient gels with higher concentrations at the base than at the top often afford a better separation. Gradient gels are always acrylamide based and can be made by manually pouring different concentrations into tubes or plates, or by use of a gradient maker (*Figure 3.4*).

Agarose is also used, not just to strengthen gels but also for preparing agarose gels which have a sieving matrix similar to acrylamide

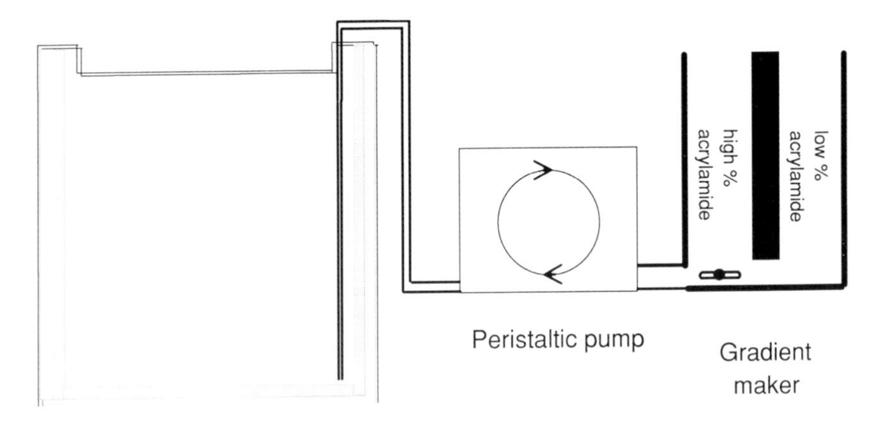

Figure 3.4: A simple gradient maker for the preparation of gradient gels.

Table 3.4: Gel buffers

Tris-borate (1 × TBE)
10.8 g Tris base (89 mM)
5.50 g boric acid (89 mM)
0.93 g $EDTA.Na_2.2H_20$ (2.5 mM)
1.00 g SDS (0.1%(w/v))
made up to 1 liter with distilled water
Note: boric acid adjusts pH to 8.3 at 25°C

Tris acetate (1 × TAE)
4.84 g Tris base (40 mM)
0.372 g $EDTA.Na_2.2H_20$ (1 mM)
pH adjusted to 8.0 with glacial acetic acid

Tris-phosphate
0.44 g Tris base (3.6 mM)
0.48 g $NaH_2PO_4.2H_20$ (3.0 mM)
0.04 g $EDTA.Na_2.2H_20$ (0.1 mM)
1.00 g SDS (0.1%(w/v))
made up to 1 liter with distilled water
Note: $NaH_2PO_4.2H_2O$ adjusts pH to 7.7 at 25°C

matrices. These gels are also used for separating RNA and have the advantage that they are much easier to prepare. The agarose is melted into the buffer of choice (*Table 3.4*) to a final concentration of between 1 and 1.6% (w/v), and is usually poured as a slab gel. For example, mRNA is usually analyzed by Northern blotting in 1–1.5% agarose gels. The choice of gel matrix and gel concentration again depends on the RNA being separated and the conditions to be used during separation.

3.3.2 Factors influencing molecular migration

The basis of the separation of RNA by gel electrophoresis has much in common with centrifugal separations, in that molecules of RNA are separated on the basis of their size and conformation. Molecular size is the main determinant of migration speed where smaller molecules move through the matrix faster than larger molecules. Compact molecules also migrate faster than extended molecules of the same size. However, the same problems of RNA aggregation, as found in centrifugal preparations, are also encountered. It is therefore necessary to use denaturing gels to determine the actual size of RNA in the absence of any conformational factors, aggregation and nicks in the RNA. The choice of denaturing conditions depends on the nature of the gel and the electrophoresis conditions; some of the denaturants used are given in *Table 3.5*. It should be noted that this method is best suited to linear RNA assessment; non-linear (i.e. branched) RNA migrates more slowly than would be expected from its size as it cannot pass through the pores as easily.

Table 3.5: Types of denaturing conditions used for gel electrophoresis

Denaturant	Concentration	Type of gel	Comments
DMSO	50–90%	Agarose	Toxic
Glyoxal	10–30%	Agarose	Usually used in combination with DMSO in denaturing agarose gel electrophoresis
Formaldehyde	3% (w/v)	Agarose and polyacrylamide	Toxic. Usually used in denaturing agarose gel electrophoresis
Formamide	50–98%	Agarose and polyacrylamide	Toxic. Usually used in loading buffers
Heat	60–80°C	Agarose and polyacrylamide	Usually used in concert with other denaturants for initial denaturation
Methyl mercuric hydroxide	3–5 mM	Agarose	Highly toxic. Reacts with imino groups of uridine and guanosine in RNA preventing secondary structure formation. Inhibits polymerization of polyacrylamide gels
Sodium iodoacetate	10 mM	Agarose	Used in the denaturing glyoxal/DMSO system for Northern blotting
Urea	6–8 M	Polyacrylamide	Used in denaturing PAGE protocols. Inhibits formation of agarose gels

3.3.3 Polyacrylamide slab gel electrophoresis

Non-denaturing polyacrylamide gels. Polyacrylamide slab gels are frequently used for RNA separation. Eight or more samples can be separated in the same gel matrix. The use of slab gels allows samples to be aligned and compared more easily and accurately.

Protocol 3.1: Polyacrylamide gel electrophoresis

1. The electrophoresis tank (*Figure 3.5*) has upper and lower reservoirs which are connected only when a gel is in position. This ensures that any electrophoretic flow passes through the gel and not through the running buffer.
2. Gels are cast between two glass plates separated by 1–3 mm thick plastic strips. Plates are cleaned before each use with laboratory detergent, rinsed with distilled water and then with a 1:1 (v/v) ethanol–ether solution and dried with tissue.
3. Plates are sealed with rubber tubing and 3% (w/v) agarose before gel mixes are poured (*Figure 3.6*).
4. An appropriate gel mixture (*Table 3.3*) is prepared and degassed for 10 min to remove oxygen that inhibits polymerization.
5. Cross-linking is initiated by addition of 25 µl of the catalyst TEMED and 250 µl freshly dissolved 10% (w/v) ammonium persulfate (APS).
6. The gel mixture is slowly poured down the edge of the space between the plates.
7. Gels usually set in under 30 min but it is normal procedure to leave them overnight to ensure complete polymerization.
8. Samples are loaded with 1/10 vol. load buffer (electrophoresis buffer supplemented with 10–20% (w/v) sucrose, 0.25% (w/v) bromophenol blue, 0.25% (w/v) xylene cyanol FF). As a guide, the maximum amount of RNA that can be loaded on to a gel is

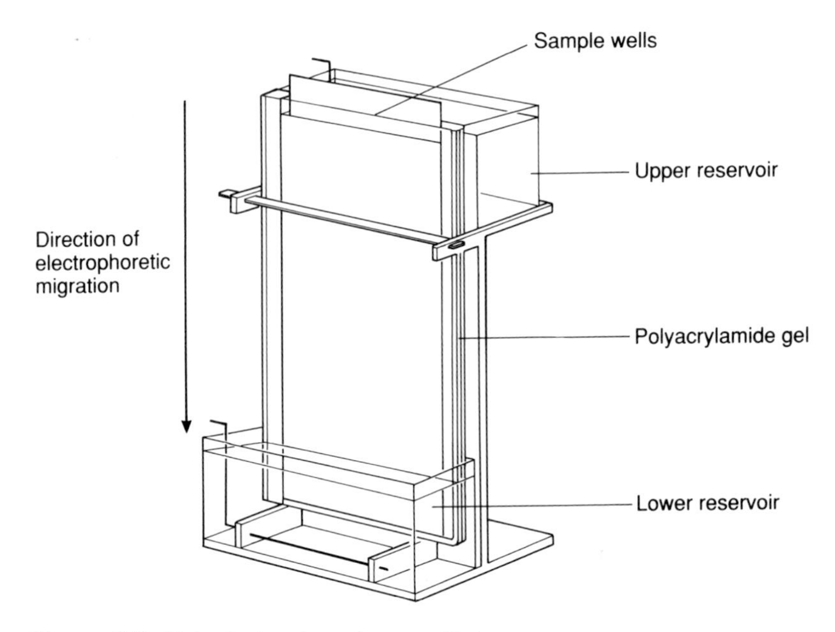

Figure 3.5: Slab electrophoresis apparatus.

about 1 μg sample RNA/μl; the minimum amount of sample depends on the method of detection.

9. The optimum electrophoretic conditions depend on the gel size, concentration and thickness. A standard homogeneous non-denaturing 7.5% polyacrylamide gel 18 × 18 × 0.2 cm can be run at 50 V for 4–5 h in a suitable running buffer (*Table 3.4*).
10. RNA can then be visualized as described in Section 3.4.

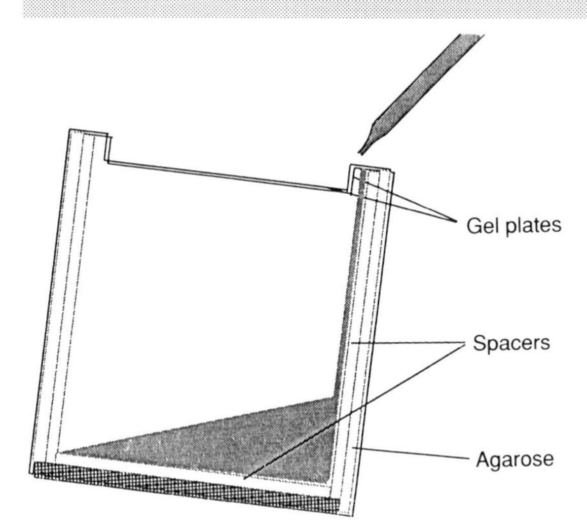

Figure 3.6: Gel mold for preparing slab gels.

Table 3.6: Gel solutions for denaturing gels

System	Gel constituents	Comments
Formamide	4% acrylamide: 0.91 g acrylamide, 0.09 g bisacrylamide, 92 mg diethylbarbituric acid, 20 ml deionized formamide and 60 μl TEMED. Mixture adjusted to pH 9.0 with concentrated HCl and made up to a final volume of 25 ml with deionized formamide. Polymerized by addition of 200 μl of 20% APS	Acrylamide and formamide are toxic, avoid contact and inhalation. Diethylbarbituric acid can be replaced with 10 mM NaH_2PO_4 (pH 6.0) or 10 mM NaH_2PO_4 (pH 9.0)
Urea	4% acrylamide: 5 ml stock acrylamide, 3.75 ml gel buffer, 9.7 ml 8 M urea. Stock acrylamide: 15 g acrylamide, 0.75 g bisacrylamide made up to 100 ml with 8 M urea. Gel buffer: see Table 3.8. Polymerized by addition of 25 μl TEMED and 200 μl fresh APS	Acrylamide and formamide are toxic; avoid contact and inhalation

Denaturing polyacrylamide gel electrophoresis. Gel mixes for denaturing polyacrylamide gels must include a denaturant to maintain the denatured state of the RNA loaded into the gel. *Table 3.6* outlines gel solutions used in the formamide- and urea-based denaturation systems.

Samples for denaturing gel electrophoresis are generally dried *in vacuo*, dissolved in a suitable volume of denaturing sample buffer and heated to ensure complete denaturation before loading. *Table 3.7* details two common sample buffers used in the two standard denaturing techniques. The amount of RNA loaded on to polyacrylamide gels varies greatly. As little as 10 ng/band can be visualized using ethidium bromide staining and as much as 5 μg/band can be viewed in isolation from other bands. Optimal results are therefore often achieved by loading 20–30 μg RNA in a volume of 10–20 μl.

Denaturing gels are run for similar lengths of time but sometimes need to be run in an incubator at high temperatures. Running buffers used for the two common denaturation systems are provided in *Table 3.8*.

Table 3.7: Sample buffers used for denaturing gels

System	Sample buffer	Incubation temperature and time	Comments
Formamide	100% formamide containing 10% sucrose	65°C for 2 min	Formamide is toxic; avoid contact and inhalation
Urea	8 M urea and 10% sucrose	60°C for 1 min	

Table 3.8: Running buffers for denaturing gel electrophoresis

System	Running buffer	Comments
Formamide	1. 20 mM NaCl in water when diethylbarbituric acid is used in the gel mix or 2. 10 mM sodium phosphate (pH 6.0 or 9.0) in water when 10 mM NaH_2PO_4 (pH 6.0) or 10 mM Na_2HPO_4 (pH 9.0) are used in the gel mix or 3. 98% deionized formamide 4. 150 mM sodium phosphate buffer, 5 mM EDTA, 180 mM Tris base made up to 1 liter with 8 M urea[a]	Formamide is toxic; avoid contact and inhalation
Urea	Standard buffers (Table 3.4) may be used at 60°C and/or buffer made up in 8 M urea	Electrophoresis is carried out at 60°C

[a]23.8 g $NaH_2PO_4.2H_20$ (150 mM), 1.85 g $EDTA.Na_2.2H_20$ (5 mM), 1.7 g Tris base (180 mM) pH 7.7 made up to 1 liter with 8 M urea.

3.3.4 Agarose gel electrophoresis

Submerged or 'submarine' slab gels in which the gel is completely immersed (*Figure 3.7*) are popular because they are easy to prepare, load and analyze [5]. Agarose slab gels allow separation of 10 or more RNA samples on the same gel with only a slightly lower resolution

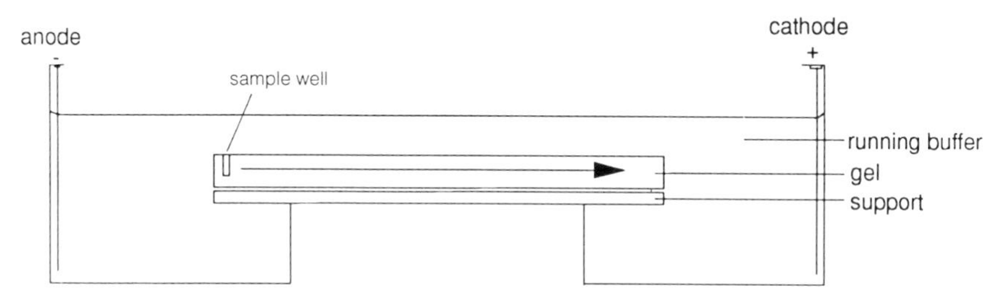

Figure 3.7: Submarine slab gel apparatus.

than polyacrylamide slab gels. The advantages of the use agarose far outweigh this slight loss of resolution. Agarose is a non-toxic compound, less expensive than polyacrylamide and needs no cross-linking agents or polymerization agents.

Protocol 3.2: Agarose gel electrophoresis

1. Agarose gels are made by melting the agarose powder in a volume of water or buffer for denaturing and non-denaturing gels, respectively. Agarose can be melted in a water bath or in a microwave oven. *Table 3.9* gives some buffers used for preparing non-denaturing and denaturing agarose gels. The concentration of agarose used depends on the size of the RNA to be separated. For example, mRNA typically ranges from 500 bases to 5 kb in size. For RNA larger than 1 kb, agarose concentrations of 0.8–1.0% should be used, while for RNA smaller than 1 kb the more standard concentration of 1.5% is used.
2. The molten gel solution is poured into a gel tray that has been sealed with plastic tape (*Figure 3.8*) and allowed to set. Gels containing formamide should be allowed to set at 5–10°C whilst other types of gel can be set at room temperature.
3. Gels should be left for 1 h after the gel has started to set to ensure complete matrix formation. This period should not be extended for much longer as water evaporates from agarose gels, resulting in gel shrinkage.
4. Samples are denatured before loading, with solutions appropriate to the system being used. *Table 3.10* lists some common sample buffers. The role of these solutions is to ensure that samples are totally denatured before they enter the gel matrix. This is often aided by the use of high temperatures for various times.
5. Note that RNA molecular weight markers ('ladders') must be treated in exactly the same manner as the samples, being subject to pre-electrophoretic denaturation, loading and electrophoresis.
6. Samples at a concentration of about 1 μg RNA/μl are loaded with 1/10 sample volume of loading buffer (*Table 3.11*) into the

wells. The purpose of these loading buffers is to raise the sample density so that they can be loaded as narrow layers in the sample wells of the gel and to provide a visualization of the extent of electrophoretic migration.

7. Electrophoresis is carried out for 2–3 h at 80 mA/100 V in a suitable running buffer (*Table 3.12*).

Table 3.9: Buffers used for agarose gels

Buffer	Gel concentration (w/v)	Nature	Comments
Tris-borate (TBE) (*Table 3.4*)	0.8–1.6%	Non-denaturing	1.5% gels used for assessment of RNA integrity
5 mM methyl mercuric hydroxide in a TBE buffer[a]	0.8–1.5%	Denaturing	1.5% is standard. Highly poisonous, volatile chemical
DMSO/glyoxal system uses 0.01 M sodium phosphate for gel preparation	0.8–1.5%	Non-denaturing	1.5% is standard, used for Northern blotting of RNA. Buffer needs to be recirculated. This system affords sharper banding than formaldehyde based systems but is more difficult to run. Sodium iodoacetate can be added to inactivate ribonucleases
MOPS[b]: 1 × MOPS buffer[c] 6.5% formaldehyde	0.8–1.5%	Denaturing	1.5% is standard, used for Northern blotting of RNA

[a]Gel buffer: 50 mM boric acid, 10 mM Na_2SO_4, 5 mM disodium tetraborate, 1 mM EDTA.
[b]MOPS, 3-(*N*-morphino)propanesulfonic acid.
[c]10 × MOPS buffer (stock concentration): 0.2 M MOPS, 50 mM sodium acetate, 5 mM EDTA.

3.3.5 Two-dimensional gel electrophoresis

Two-dimensional electrophoresis is used to separate complex nucleic acid mixtures that cannot be resolved in a one-dimensional gel. Polyacrylamide gels are used in this method, where a shift in electrophoretic conditions occurs between the first and second dimensions. The shift in separation conditions can be achieved by a polyacrylamide concentration change, a shift from non-denaturing to denaturing conditions or a shift in pH combined with denaturation and concentration changes.

A shift in polyacrylamide concentration allows RNA molecules with different conformations to be separated. Here, the acrylamide concentration in the second dimension is twice that of the first dimension. A concentration shift of this type often gives well-defined, tight banding and can be used to separate RNAs that vary in size between 80 and 400 nucleotides [7].

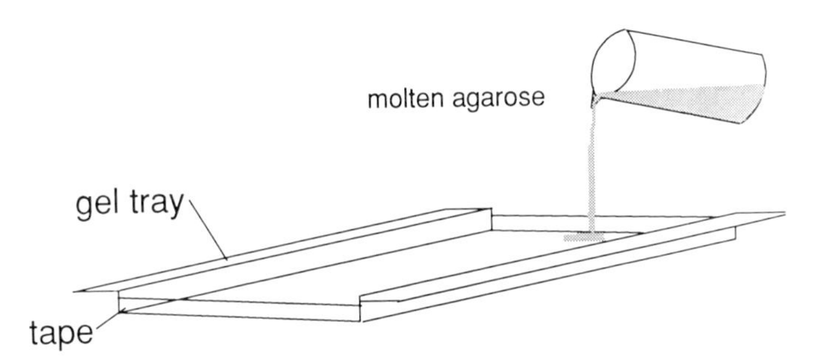

Figure 3.8: Preparation of a mold tray for agarose gels using plastic tape.

Table 3.10: Sample buffers used for denaturing agarose gel electrophoresis systems

System	Sample buffer	Incubation temperature and time	Comments
Methyl mercuric hydroxide	12.5 mM methylmercuric hydroxide, 50 mM boric acid, 5 mM disodium tetraborate, 10 mM Na_2SO_4, 10% (w/v) glycerol and 0.1% w/v bromophenol blue	None	Methylmercuric hydroxide is volatile and highly toxic
MOPS/ formaldehyde	0.7 × MOPS buffer, 9.2% formaldehyde	65°C for 2 min	Formaldehyde is toxic; avoid contact and inhalation
Glyoxal/DMSO	1 M Glyoxal, 50% DMSO, 0.01 M sodium phosphate	50°C for 60 min or 65°C for 15 min	Toxic

Table 3.11: Loading buffers used in selected agarose gel electrophoresis systems

System	Load buffer	System nature	Comments
Tris-borate (TBE)	50% glycerol, 1 mM EDTA (pH 8.0), 0.25% bromophenol blue, 0.25% xylene cyanol FF	Non-denaturing	Common system for quick tests of RNA integrity
Methylmercuric hydroxide	None—sample buffer contains glycerol and bromophenol blue	Denaturing	Highly toxic, volatile methylmercuric hydroxide should remain in the gel
MOPS/ formaldehyde	50% glycerol, 1 mM EDTA (pH 8.0), 0.25% bromophenol blue, 0.25% xylene cyanol FF	Denaturing	
Glyoxal/DMSO	50% glycerol, 10 mM sodium phosphate (pH 7.0), 0.25% bromophenol blue, 0.25% xylene cyanol FF	Denaturing	

Table 3.12: Running buffers used for denaturing agarose gel electrophoresis

System	Running buffer	Comments
Methylmercuric hydroxide	50 mM boric acid, 10 mM Na_2SO_4, 5 mM disodium tetraborate, 1 mM EDTA	No methylmercuric hydroxide is required in the running buffer as it remains in the gel
MOPS/ formaldehyde	1 × MOPS buffer, 3% (v/v) formaldehyde	Formaldehyde is toxic; avoid contact with skin and eyes and inhalation, particularly whilst loading gel
Glyoxal/DMSO	10 mM sodium phosphate	Non-toxic running buffer

A shift between non-denaturing and denaturing conditions separates molecules that have hidden nicks in their structure. Urea is used as the denaturation agent in these shift experiments at concentrations varying from 4 to 8 M. RNA molecules containing nicks can remain as single units due to the secondary and tertiary interactions. In denaturing conditions these interactions are prevented and separation occurs.

A change of pH from acidic pH 3.3 in a denaturing gel to non-denaturing conditions in higher acrylamide concentration gel at pH of 8.0–8.3 forms the basis of the third method. Here, initial separation in an acidic environment separates the molecules on a basis of base composition. Nucleotide bases (A, C, G, U) contribute different charges to the net charge of a RNA molecule at pH 3.3. Therefore, molecules of discrete compositions have different mobilities in electric fields under these conditions. This separation factor is rarely sufficient to differentiate between molecules in isolation but can be combined with concentration and denaturation shifts [8].

The first and second dimensions are physically separate. Samples are electrophoresed through a tube or slab gel that constitutes the first dimension. The first-dimensional gel is placed into a gel mold around which the second-dimensional gel is poured (*Figure 3.9*). Electrophoresis

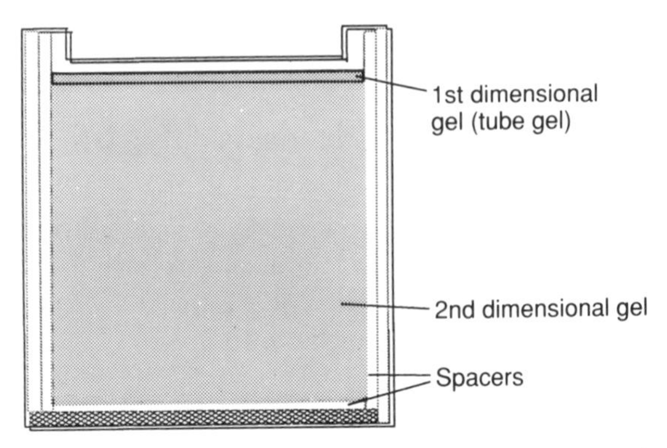

Figure 3.9: Preparation of the second-dimensional gel for two-dimensional gel electrophoresis.

in the second dimension allows fragments separated in the excised band, to migrate under different conditions through the second gel.

Gel concentrations are defined by the sizes of RNA to be separated. For example, mRNA varying in size from 800 bases to 5 kb can be separated on a 6% polyacrylamide gel whilst smaller tRNA can be separated on a 15% polyacrylamide gel.

Examples of the three main types of two-dimensional gel separations are described in *Table 3.13*.

Table 3.13: Selected two-dimensional systems used for separating RNA molecules

Shift type	**First dimension**		**Second dimension**			
	Gel concentration	**Conditions**	**Gel concentration**	**Conditions**	**Type of RNA separated**	**Refs**
Polyacrylamide concentration	10%	Non-denaturing at neutral pH	20%	Non-denaturing at neutral pH	small RNAs	9
	10.4%	Denaturing (4 M urea) at neutral pH	20.8%	Denaturing (4 M urea) at neutral pH	tRNAs and precursors	10
Denaturation	6%	Non-denaturing at neutral pH	6%	Denaturing (5 M urea) at neutral pH	mRNAs	11
	12.5%	Non-denaturing at neutral pH	12.5%	Denaturing (8 M urea) at neutral pH	5S RNA fragments	12
	15–16%	Denaturing (6–7 M urea) at neutral pH	16%	Non-denaturing at neutral pH	tRNAs	13
pH, denaturation and polyacrylamide concentration	10.3%	pH <4.5 in denaturing conditions (6 M urea)	20.6%	Non-denaturing at neutral pH	Viral RNA fragments	14
	10.3%	pH <4.5 in denaturing conditions (6 M urea)	20.6%	Non-denaturing at neutral pH	Viral RNA oligonucleotides	15

3.4 Analysis of gel electrophoretic separations

The choice of analytical method must be integrated with the needs of the experiment. Preparative methods must incorporate a visualization step which will not affect RNA structure and function, required for later experimental work. Where analysis is the end-point of an experiment, methods that modify the RNA can be used.

3.4.1 UV shadowing and UV scanning

Frequently UV shadowing can be used to indicate the position of RNA bands within the gel. This is done by placing the gel on clear polythene above a Polygram Cel 300 UV 234 thin-layer chromatography sheet (Macherey-Nagel). The RNA bands, which appear as shadows, can then be located under UV radiation, the gel segment cut out and the RNA eluted. Gels can also be scanned with UV scanners; tube gels need to be removed from their tubes before scanning unless they have been run in quartz tubes. A resolution of as little as 0.1 µg may be seen as a peak above background. Peak area is also approximately proportional to RNA concentration.

3.4.2 Staining methods

Direct staining of the bands gives an approximate estimate as to RNA concentration and often more importantly is used to check that the RNA is not degraded. A simple method for RNA analysis is ethidium bromide staining. Ethidium bromide (EtBr) has a planar structure and as such can intercalate between bases. Intercalation of EtBr into single-stranded nucleic acid occurs at a lower frequency than into double-stranded nucleic acid. However, RNA is still easily visualized using this method. Ultraviolet (UV) radiation boxes are used to excite fluorescence in the RNA bands which emit an orange light (590 nm). The intercalated form of ethidium bromide has a much higher potential for excitation than free EtBr molecules. This produces intensely pink stained bands and a faintly glowing background. A camera system mounted above the transilluminator can be used to take photographs of gels. For photographing gels an orange filter is usually placed in front of the lens and typical settings of f4.5, 1/30–1/4 sec with Polaroid film are used. Recently, video cameras mounted above transilluminators have allowed image capture which can be printed on to thermal paper at a much reduced cost.

This method, although not very sophisticated, it is probably the most commonly used direct staining method. Gels can also be stained with methylene blue or pyronin Y (*Table 3.14*) after fixing in 1 M acetic acid or 10% trichloroacetic acid for 30–60 min.

Slab gels are not usually stained in their entirety, instead the outer lanes, loaded with commercial RNA size markers are cut from the gel and stained. This allows assessment of size without interfering with the sample RNA. The remaining samples are subsequently analyzed by, for example, Northern blotting. *Table 3.15* lists some commercially available RNA molecular markers.

A hard copy record of the location of each band allows its mobility to be measured. RNA of the required size can then be located and eluted or the information can be used to size radiolabeled bands on later autoradiographs. The sizes of RNA are estimated from the relative

Table 3.14: Staining agents used to visualize RNA in gels

Staining agent	Procedure	Comments
Ethidium bromide (EtBr)	Stain with 5 μg/ml ethidium bromide dissolved in 0.5 M ammonium acetate for 30–60 min and destain by washing gel in water	5 μg/ml ethidium bromide can be included in the gel matrix solution in some cases, bypassing the need to stain the gel after electrophoresis. However, this does reduce the mobility of RNA but can can give information as to RNA integrity
Methylene blue	Soak gel overnight in 0.2% methylene blue dissolved in 0.2 M sodium acetate buffer (pH 4.7). Destain by washing in water	Toluidine blue O, thionin and Azure A can be used in place of methylene blue
Methylene blue	A membrane blot of a gel can be stained after baking by soaking in 5% acetic acid for 15 min at room temperature before being transferred to 0.5 M sodium acetate (pH 5.2) and 0.04% methylene blue for 5–10 min	
Pyronin Y	Stain with 0.1% pyronin Y dissolved in 0.5% acetic acid, 1 mM citric acid and destain in 0.5% acetic acid	

Table 3.15: Commercial sources of RNA markers

Commercial description	Supplier	Band sizes (kb)	Comments
0.16–1.77 kb RNA ladder	Gibco–BRL	1.77, 1.52, 1.28, 0.78, 0.53, 0.40, 0.28, 0.155	Suitable for sizing single-stranded RNA in formaldehyde or glyoxal gels. Visualization by EtBr staining or by autoradiography (end labeling dephosphorylated ladder)
0.24–9.5 kb RNA ladder	Gibco–BRL	9.49, 7.46, 4.40, 2.37, 1.35, 0.24	Suitable for sizing single-stranded RNA in formaldehyde or glyoxal gels. Visualization by EtBr staining or by autoradiography (with nick translated λ DNA)
The radiolabeled RNA ladder system	Gibco–BRL	9.5 to 0.78	T7 RNA polymerase based system producing ^{32}P labeled bands. Detectable in gel or on a Northern blot by autoradiography
RNA markers, 0.36–9.49 kb	Promega	9.488, 6.225, 3.991, 2.8, 1.898, 0.872, 0.562, 0.363	Suitable for sizing single-stranded RNA in formaldehyde or glyoxal gels. Visualization by EtBr staining or by autoradiography (end labeling dephosphorylated ladder)
rRNA, 23S and 16S (*E. coli* R13)	Pharmacia	3.566 and 1.776	rRNA markers remain intact under denaturing conditions
rRNA, 28S and 18S (calf liver)	Pharmacia	6.333 and 2.366	rRNA markers remain intact under denaturing conditions

migration of markers of known size in much the same way as is used for isokinetic sucrose gradients. However, it is more usual to plot the linear relationship of mobility against the molecular weight, an example of such a plot is shown in *Figure 3.10*.

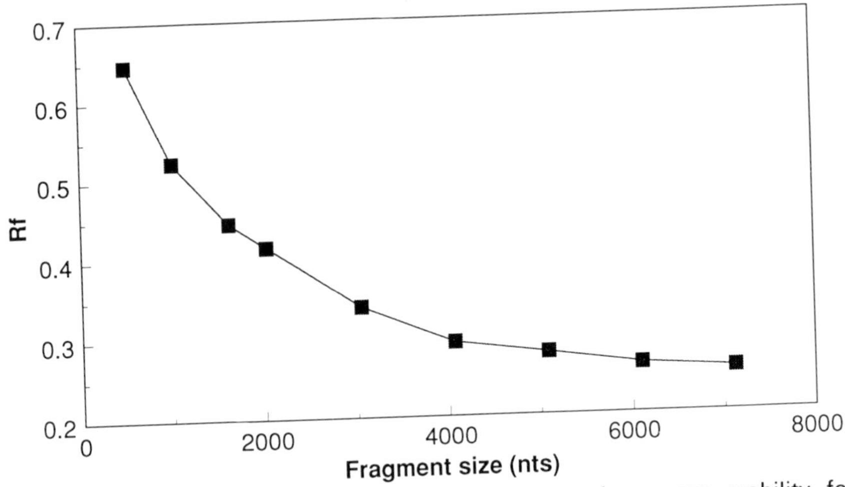

Figure 3.10: An example of a plot of fragment size versus mobility for determining the size of RNA.

3.5. Detection of specific RNA sequences

Radiolabeling is a powerful method which can define location and concentration of RNA molecules. It can be used with varying degrees of specificity from labeling many species of RNA at once, to labeling only one species. A frequent requirement is to determine the position and amount of a particular RNA species. This can be achieved by either blotting the RNA on to a membrane or if the RNA is already labeled, direct measurement is possible.

3.5.1 Radiolabeling of nucleic acids

The five basic methods for radiolabeling nucleic acid probes are described here.

Nick translation. DNA can be labeled using this method. Double-stranded DNA (dsDNA) is nicked with DNase I providing a target site for radiolabeling. DNA polymerase can repair the nick by incorporating free nucleotide triphosphates. It is usual in labeling reactions to substitute one of the nucleotides with a ^{32}P-radiolabeled deoxyribonucleotide triphosphate. DNA polymerase incorporates nucleotides with a concomitant removal of two phosphate groups from the incoming nucleotide. Therefore the nucleotides must have their α-phosphate labeled in order to label the DNA strand. DNA polymerase incorporates nucleotides by adding a free nucleotide to the 3′-OH group that is available. DNA polymerase also displaces the 5′ nucleotide of the nick as a result of its 5′→3′ exonuclease activity replacing it with a free nucleotide complementary to the template DNA strand. In this manner, DNA polymerase removes 5–20 bases replacing them with new nucleotides, some of which are radiolabeled.

Template DNA can be isolated from DNA stock plasmids or can be generated by the polymerase chain reaction (PCR) (Section 5.5). Double-stranded DNA can be isolated from a plasmid vector by restriction enzyme digestion to yield a fragment complementary to the RNA of interest. This DNA can be isolated from plasmid DNA by electrophoresis in a non-denaturing agarose slab gel (Section 3.3.4) followed by electroelution (Section 3.6.2). DNA should be purified by a phenol–chloroform–isoamyl alcohol extraction, a chloroform–isoamyl alcohol extraction and passage through Sepharose CL6B (Sigma) spin-columns to remove any contaminants that may interfere with radiolabeling.

Protocol 3.3: Preparation of Sepharose CL6B spin-columns

1. Sepharose CL6B columns may be made by adding 0.5 ml prewashed (with TE) Sepharose CL6B to a 0.5 ml microcentrifuge tube which has a small hole in the base.
2. The column is formed by centrifuging the tube in a bench-top swing-out rotor at 1500 *g* for 3 min at room temperature.
3. Between 15 and 50 µl DNA is then added to the top of the column which is then placed inside a sterile 1.5 ml microcentrifuge tube.
4. The DNA is eluted from the column by centrifugation at 1500 *g* for 3 min into the 1.5 ml microcentrifuge tube.
5. DNA concentration should be assessed by comparison of ethidium-stained intensity to that of ladder DNA of known concentration or spectrophotometrically.

Protocol 3.4: Radiolabeling using nick translation

1. Reaction mixture:
   1 µg template DNA in a maximum volume of 23 µl
   5 µl of 10 × nick translation buffer (500 mM Tris-HCl (pH 7.2), 100 mM $MgSO_4$, 1 mM dithiothreitol (DTT))
   10 µl nucleotide mix (500 µM each dNTP omitting radiolabeled type)
   7 µl [α-^{32}P]dCTP (14.8 TBq (400 Ci)/mmol and 370 MBq (10 mCi)/ml) 5 µl optimized enzyme mix (DNase I/DNA polymerase) (Promega)
   Reaction volume is made up to 50 µl with sterile, double-distilled water.

2. Incubation at 15°C for 1 h allows nicking of DNA and repair by DNA polymerase thus incorporating radiolabeled nucleotides.
3. The reaction is terminated by the addition of 5 μl stop mix (0.2 M EDTA, pH 8.0).
4. Unincorporated radiolabel can be removed by gel filtration using a Sepharose CL6B (Pharmacia) or Sephadex G-50 (Pharmacia) column.
5. The double-stranded DNA produced must be denatured before it can be used for hybridization experiments. This is usually achieved by placing the sample at 100°C for 5 min before transferring to ice (preventing reannealing).

This method usually incorporates more than 65% of the label into the DNA.

CAUTION: Heating samples to 100°C causes a rise in pressure in the microcentrifuge tube containing the sample. To avoid the possibility of spilling radioactivity upon boiling, a sealed screw-threaded tube should be used which will not pop open.

Primed synthesis. In common with nick translation, primed synthesis methods use the ability of DNA polymerase to synthesize a new DNA strand, starting from a free 3′-OH group. In this method a short oligonucleotide primer is annealed to a single-stranded DNA template (ssDNA). It is essential to use a polymerase lacking a 5′–3′ exonuclease activity to avoid degradation of the primer. Two enzymes, the Klenow fragment (the large fragment of the *E. coli* DNA polymerase holoenzyme) and reverse transcriptase both fulfil these needs. A favored method is based on primed synthesis using random hexanucleotides as primers.

In this method a mixture of random hexanucleotides is used to prime double-stranded DNA synthesis from a linear, single-stranded DNA template. Typically [α-^{32}P]dNTP are used to label synthesized strands. The sizes of the template should exceed 200 bp as smaller templates produce very small labeled oligonucleotides with low activity and a consequent high background during hybridization. The labeled fragments produced by this method are longer than those produced by nick translation and as such can be used in a more discriminating manner with respect to probing for RNA sequences.

Protocol 3.5: Random primed synthesis method [16]

1. Template DNA isolated as described earlier in this section is purified using Sepharose CL6B spin-columns.

2. The DNA concentration should be determined by comparison with ethidium bromide-stained DNA standards of known concentrations.
3. The dsDNA template in a volume of 11 μl (25–100 ng) is heat denatured at 100°C for 5 min.
4. The separate strands are prevented from reannealing by transferring the sample into ice for 3 min. This procedure called snap-cooling, prevents reannealing which would occur if the sample was to cool slowly.
5. An equal volume (11 μl) of 'master-mix' (43.8 μM dCTP, dGTP, dTTP, 0.438 M HEPES–NaOH pH 7.0, 12.3 U/ml random hexanucleotide (Pharmacia)), 0.5 μl Klenow (3 U) and 2 μl of [α-^{32}P]dATP (222 TBq/mmol, 6000 Ci/mmol NEN, Du Pont)* are added.
6. The reaction mixture is incubated at room temperature for 2–5 h to synthesize the DNA.
7. Synthesis is terminated by the addition of 5 μl stop mix (50 mM EDTA, 1 mg/ml calf thymus DNA, dextran 804, 0.1% (w/v) bromophenol blue).
8. The labeled DNA is purified from free nucleotides by passage through Sepharose CL6B spin-columns.
9. 100 μl sterile, double-distilled water is added to the labeled DNA.
10. Labeled strands are separated from the template DNA by heat denaturation at 100°C for 5 min and followed by snap-cooling in ice to prevent reannealing.
11. The probe may then be used immediately for hybridization or frozen at −20°C until needed.

*CAUTION: [α-^{32}P]dATP is a high-intensity radioactive source. Shielding with at least 8 mm Perspex at all times is imperative. Stocks of [^{32}P]-labeled nucleotide should be removed from the freezer at least 30 min before use to allow thawing and mixed well before use. Opening frozen nucleotides can create radioactive aerosols.

Primed synthesis from a specific oligonucleotide. In a similar manner to the random priming method, DNA synthesis can be initiated from an oligonucleotide of defined sequence. This allows the template to be used in a defined manner to generate specific probes. The enzymes used in primed synthesis methods do not have a high processivity, dropping off the template after copying at most 100–200 bases. Thus, short DNA stretches can be copied to give specific DNA probes that give very low background noise for autoradiography.

LIVERPOOL
JOHN MOORES UNIVERSITY
AVRIL ROBARTS LRC
TEL. 0151 231 4022

RNA polymerase-based methods for labeling RNA. RNA polymerases synthesize RNA from ribonucleoside triphosphates using a DNA template. Thus, they have the ability to incorporate radiolabeled ribonucleotides into RNA molecules which can be used as molecular probes for hybridization analyses. The labeled RNA transcript is commonly referred to as a 'riboprobe'. A variety of vectors incorporate a bacteriophage promoter which is upstream from a multiple cloning site, allowing transcription of the inserted DNA. The bacteriophage promoters are usually incorporated into plasmid vectors in a pairwise manner, placing one promoter either side of the insert. This allows transcription in either direction depending on the orientation of the gene. There is an enormous variety of plasmids available for generating labeled probes, with new commercial vectors being produced almost monthly. *Table 3.16* details some commercially available plasmid vectors suitable for cloning a desired DNA insert.

The method of cloning a DNA fragment into a vector in order to generate an RNA transcript was once the only option. However, PCR has made cloning unnecessary in many cases. PCR (Section 5.5) exponentially amplifies a target DNA or RNA (RT-PCR) sequence by a cyclical denaturing, annealing of primers and copying of template, returning then to the denaturing step with double the amount of template. The primers used to initiate DNA synthesis must be complementary to the target sequence at their 3′-('anchor') terminus. The 5′-terminus, however, does not have to be complementary to the target sequence. This has led to specific tailoring of primers to incorporate enzyme recognition sites into them. Some of these tailored

Table 3.16: Selected vectors used for *in vitro* RNA synthesis

Vector series	Examples	Bacteriophage RNA polymerase promoters	Comment
pBluescript (pBS)	pBS SK$^+$	T3 and T7	Major series of plasmid vectors used for general cloning and to derive RNA transcripts
pGEM	pGEM-3Z	SP6 and T7	13 restriction enzyme sites in the multiple cloning site. Extensive series of plasmid vectors available from Promega
pSPORT	pSPORT1	SP6 and T7	19 restriction enzyme sites in the multiple cloning site. Available from Gibco–BRL
pT7T3	pT7T3 18U	T3 and T7	T3 and T7 promoters flank pUC 18 multiple cloning site. Available from Pharmacia

sites include bacteriophage RNA polymerase recognition sites. Therefore DNA can be produced by use of tailored primers in PCR reactions which incorporate a proximal RNA polymerase binding site. Adding the appropriate RNA polymerase to an aliquot of a PCR reaction will initiate RNA synthesis.

RNA transcripts must be generated from the insert only and not the plasmid DNA in order to avoid later hybridization to plasmid fragments sometimes present. Therefore, before transcription is carried out the vector must be linearized. The purity of the DNA has little effect on subsequent transcription making it possible to use plasmid DNA minipreps [4,17] followed by phenol–chloroform extraction to remove the proteins.

Protocol 3.6: Synthesis of 'riboprobes'

1. The plasmid is linearized by use of restriction enzymes that cut at unique sites located after the insert. The restriction enzymes favored for linearization generate 5′ protruding termini or blunt-ended DNA. The 3′ protruding termini generated by restriction nucleases must be 'blunted' using the Klenow fragment of DNA polymerase I or T4 DNA polymerase. The amount of digestion should be checked using ethidium bromide gels (Section 3.4.2) to ensure that digestion has proceeded to completion.
2. Linearized DNA should be extracted with equal volumes of phenol–chloroform–isoamyl alcohol (25:24:1) to remove the enzymes used in linearization and blunting or PCR procedures.
3. Transcription mixture:
   1.0 μl 40 mM Tris-HCl (pH 7.5), 5 mM NaCl, 6 mM $MgCl_2$, 2 mM spermidine-HCl
   0.1 μl 1.0 M dithiothreitol (DTT)
   0.5 μl 100 μg/ml bovine serum albumin (BSA) (Fraction V; Sigma)
   1.0 μl 500 μM of each rCTP, rGTP, rUTP
   1.0 μl 20 nM linearized plasmid DNA
   5.0 μl 2–20 μM [α-^{32}P]rATP (specific activity of 14.8–111 TBq (400–3000 Ci)/mmol) (NEN, Du Pont)
   0.5 μl 1 U/μl placental RNase inhibitor
   1.0 μl 500 U/ml bacteriophage DNA-dependent RNA polymerase
   0.4 μl autoclaved, double-distilled water
   The reaction mixture, total volume 10.5 μl, is incubated for 1–2 h at 37°C (T3 and T7 DNA-dependent RNA polymerases) or 40°C (SP6 DNA-dependent RNA polymerase).

The choice of radioisotopically labeled 5′[α-^{32}P]rNTP used in this and other labeling procedures varies between laboratories often due purely to personal preferences. Guidelines for optimal specific activities of different radioisotopes are detailed in *Table 3.17*.

Table 3.17: Specific activities of rNTPs recommended for radioisotopic labeling of RNA *in vitro*

Nucleotide	Recommended radioactivity used per reaction	Specific activity	Final concentration
[α-^{3}H]UTP	0.925 MBq (25 μCi)	1.48 TBq (40 Ci)/mmol	31 μM
[α-^{32}P]CTP	1.85 MBq (50 μCi)	14.8 TBq (400 Ci)/mmol	6 μM
[α-^{35}S]UTP	11.1 MBq (300 μCi)	48.1 TBq (1300 Ci)/mmol	12 μM

5′-End labeling

Methods for 5′-end labeling DNA. End labeling of nucleic acids usually involves the addition of a radiolabeled phosphate group on to a nucleic acid strand that is, or has been made capable of being phosphorylated. DNA strands are usually already 5′-phosphorylated which prevents immediate 5′-labeling by phosphorylation. It is possible to label the 5′ or the 3′ terminus of DNA depending on the needs of the experiment. The efficiency of 5′-end labeling is, however, greater than that achieved by 3′-end labeling.

DNA can be 5′-end labeled with [γ-^{32}P]ATP using polynucleotide kinase or 3′-end labeled using the Klenow fragment or terminal transferase. RNA can be labeled using either poly(A) polymerase (3′-end labeling) or polynucleotide kinase/guanyl transferase (5′-end labeling).

Nucleic acids usually have phosphorylated 5′ termini which must be dephosphorylated before they can be end labeled. This dephosphorylation is accomplished by use of alkaline phosphatase. Commercial kits supplying dephosphorylation and labeling enzymes and enzyme buffers are available (e.g. Promega).

The DNA used in the dephosphorylation reaction should be linearized. The concentration of template DNA to be used is determined by the number of sites to be phosphorylated (5′ termini). A target of 10 pmol of available termini is used to define the amount of fragment DNA to be dephosphorylated. The amount of fragment DNA used for dephosphorylation and labeling therefore varies with fragment size. An equation can be used to calculate the amount of DNA to be used in a labeling reaction. Based on the moles of termini to be labeled (M), the average molecular weight of the terminal nucleotide (MW), size of fragment to be labeled (S) and the number

LIVERPOOL JOHN MOORES UNIVERSITY LEARNING SERVICES

of termini to be labeled (N), then the mass of the fragment ($MASS$) required is given by:

$$MASS = M \times MW \times S/N$$

e.g. 10 pmol termini (1×10^{-11}mol) $\times$ 300 Da $\times$ 1000 bp/2 (dsDNA) = 1.5 μg fragment.

Fragment concentrations may be assessed spectrophotometrically (10 μg/ml has OD_{260} of 0.225) (Section 5.1) or electrophoretically (by comparison of fluorescence of ethidium bromide-stained DNA standards of known concentration) (Section 5.2).

Protocol 3.7: Dephosphorylation

1. Reaction mixture:
   5 μl dephosphorylation buffer (500 mM Tris-HCl, pH 9.0, 10 mM $MgCl_2$, 1 mM $ZnCl_2$, 10 mM spermidine)
   1 μl substrate DNA (up to 10 pmol 5′ ends)
   0.5 μl calf alkaline intestinal phosphatase (CAIP) (Boehringer Mannheim) adjusted to 50 μl with sterile double-distilled water.
2. The nature of the DNA termini defines the reaction time used in the dephosphorylation. If the fragment has protruding 5′ termini, as a result of restriction digestion, then reaction proceeds at 37°C for 30 min before adding a further 0.1 U CAIP and incubating for a further 30 min at 37°C. Recessed 5′ termini or blunt-ended molecules are incubated at 37°C for 15 min and 56°C for 15 min before incubating at 37°C for 15 min and 56°C for 15 min.
3. The reaction is terminated by extracting the proteins from the reaction mixture with an equal volume of phenol–chloroform–isoamyl alcohol (25:24:1). Extraction is achieved by vortexing for 1 min and centrifugation at 12 000 g for 2 min at room temperature.
4. Phenol is removed from the recovered aqueous fraction by extracting it with chloroform–isoamyl alcohol (24:1).
5. The DNA from the recovered aqueous fraction is precipitated by addition of 0.1 vol. 2 M NaCl and 2 vol. ethanol, placing at −20°C for 30 min.
6. The DNA is pelleted by centrifugation at 12 000 g for 5 min at 4°C and the supernatant is removed by aspiration.
7. The DNA may then be resuspended in 34 μl of forward exchange buffer (500 mM Tris-HCl, pH 9.5, 100 mM $MgCl_2$, 50 mM DTT, 1 mM spermidine).

LIVERPOOL JOHN MOORES UNIVERSITY
LEARNING SERVICES

Protocol 3.8: Phosphorylation

1. The reaction mixture is then added:
   15 µl [γ-^{32}P]ATP (111 TBq (3000 Ci)/mmol at 370 MBq (10 mCi)/ml)
   1 µl (8–10 U) T4 polynucleotide kinase
   Note: the ATP concentration used should be at least 1 µM.
2. DNA templates are labeled at 37°C for 10 min.
3. The reaction is terminated by addition of 2 µl 0.5 M EDTA.
4. The reaction mixture is then deproteinized by addition of an equal volume of phenol–chloroform–isoamyl alcohol (25:24:1), vortexing for 1 min and centrifugation for 2 min at 12 000 *g* at room temperature.
5. The DNA in the recovered aqueous fraction is precipitated by the addition of 0.5 vol. 7.5 M ammonium acetate and 2 vol. ethanol at −20°C for 30 min.
6. DNA can then be pelleted by centrifugation at 12 000 *g* for 5 min at 4°C and redissolved in 50 µl TE buffer (10 mM Tris-HCl, pH 7.5, 1 mM EDTA).

Methods for the 5′-end labeling of uncapped RNA. The majority of cellular RNA is uncapped (rRNA and tRNA) and can be immediately dephosphorylated and labeled. For many procedures it is necessary to calculate the amount of RNA to be used in the reaction mixtures.

The mass of the fragment required is given by:

$$MASS = M \times MW \times S/N$$

e.g. 10 pmol termini (1×10^{-11}mol) × 300 Da × 1000 bp/1 (ssRNA) = 3 µg fragment.

Protocol 3.9: Dephosphorylation

1. RNA is heat-denatured by mixing 100 pmol RNA in a 7 µl volume with 1 µl 1 M Tris-HCl (pH 8.0) at 100°C for 2 min followed by snap-cooling in ice.
2. Centrifugation for a few seconds collects the reaction mixture at the bottom of the microcentrifuge tube.
3. Dephosphorylation is achieved by addition of 1 µl CAIP (20 mU) and incubation at 50°C for 1 h.
4. The reaction is terminated and the enzyme inactivated by addition of 3 µl 50 mM nitrilotriacetic acid (NTA) (pH 7.2), followed by incubation at 50°C for 20 min.

Protocol 3.10: Phosphorylation

1. The RNA dissolved in double-distilled water is heat-denatured

by heating to 100°C for 2 min before snap-cooling on ice. The RNA solution is collected at the base of the microcentrifuge tube by brief centrifugation.
2. RNA (12 μl) is labeled by addition of 1 μl 100 mM DTT, 1 μl 0.2 M $MgCl_2$, 32 mM spermidine (Sigma), 1 μl 5′ [γ-^{32}P]ATP and 1 μl (4 U) polynucleotide kinase. The mixture is incubated at 37°C for 30 min.
3. The reaction is terminated and the RNA precipitated by addition of 16 μl 4 M ammonium acetate (pH 7.0) and 80 μl ethanol at −20°C overnight. Unincorporated 5′ [γ-^{32}P]ATP remains in the supernatant and can be removed from the labeled RNA by passage through a spin-column (Section 3.5.1, Nick translation).

As an alternative to dephosphorylation/rephosphorylation labeling procedures, RNA can also be labeled by phosphate exchange. This procedure can be used with RNA molecules possessing a 5′-monophosphate group which is transferred to ADP and replaced with 5′-[γ-^{32}P] group of a radiolabeled ATP molecule during phosphate exchange.

Protocol 3.11: Labeling by phosphate exchange

1. Reaction mix:
   100 pmol 5′-monophosphate termini
   2 μl imidazole buffer (25 mM imidazole-HCl (pH 6.6), 25 mM DTT, 0.5 M spermidine, 50 mM $MgCl_2$, 0.5 mM Na_2EDTA)
   2.5 μl 0.25 mM ADP
   1.0 μl 5′ [γ-^{32}P]ATP
   1.0 μl (4 U) polynucleotide kinase
   Volume is adjusted to 13 μl with autoclaved, double-distilled water.
2. Incubation at 37°C for 30 min precedes termination of the reaction by ethanol precipitation (9 μl 4 M ammonium acetate (pH 7.0) and 45 μl ethanol at −20°C overnight).

Uncapped RNA can be recapped with radioactively labeled nucleotides. Primary transcripts (uncapped) can also be capped using this procedure as they have the necessary 5′-triphosphate group. The enzyme that catalyzes this reaction is guanylyl transferase.

Protocol 3.12: Labeling by capping

1. RNA (100 pmol termini) is heat-denatured at 100°C for 2 min before being snap-cooled on ice.
2. The RNA solution (3.0 μl) is collected at the base of the

microcentrifuge tube by centrifugation for 20 sec in a microcentrifuge at room temperature.

3. The following reagents are added to the RNA for the capping reaction:
   1.0 µl 60 mM $MgCl_2$
   1.0 µl 100 mM DTT
   3.0 µl 5′-[α-^{32}P]GTP (1.11 MBq (30 µCi))
   1.0 µl (1 U) guanylyl transferase
   Incubation at 37°C for 90 min allows capping of RNA.
4. The reaction is terminated and RNA precipitated by addition of 10 µl 4 M ammonium acetate (pH 7.0) and 50 µl ethanol at −20°C overnight.
5. Centrifugation at 12 000 *g* for 10 min at 4°C pellets the RNA. Unincorporated 5′ [γ-^{32}P]ATP remains in the supernatant and can be removed from the labeled RNA.

Methods for the 5′-end labeling of capped RNA. The 5′ termini of most RNA molecules are phosphorylated; however, eukaryotic mRNA (apart from intramitochondrial mRNA) has a cap structure which must be considered when labeling the 5′ terminus. The 5′ termini of other RNA types may be directly labeled as described later in this section.

Decapping can be achieved chemically or enzymatically. The amount of RNA to be decapped and subsequently end labeled can be calculated as described previously. The chemical decapping procedure is based upon modification of the cap structure making it labile to aniline cleavage.

Protocol 3.13: Chemical decapping of RNA [18]

1. To 100 pmol RNA termini in a 10 µl volume, add 10 µl 20 mM $NaIO_4$ and incubate at room temperature in the dark for 60 min. This modifies the cap structure nucleoside by oxidation.
2. The reaction is terminated by precipitation of the RNA by the addition of 2 µl 2 M sodium acetate (pH 5.0) and 50 µl ethanol. The RNA is left to precipitate at −20°C overnight.
3. The RNA is pelleted by centrifugation at 12 000 *g* for 10 min at 4°C. Two further precipitations and a wash with ethanol remove remaining contaminants. The pellet is then vacuum dried.
4. Removal of the cap nucleoside is achieved by addition of 20 µl 165 mM aniline, 75 mM sodium acetate (pH 4.5) and incubation at room temperature for 4 h in the dark.
5. The reaction is terminated by precipitation of the RNA as in step 2.
6. The RNA is pelleted by centrifugation at 12 000 *g* for 10 min at 4°C. Two further precipitations and a wash with ethanol removes contaminants. The pellet is then vacuum dried.

The enzymatic decapping procedure relies on removal of the cap nucleoside by tobacco acid pyrophosphatase followed by subsequent dephosphorylation by alkaline phosphatase.

Protocol 3.14: Enzymatic decapping of RNA

1. 100 pmol RNA in a volume of 7 μl is added to 1 μl 0.5 M sodium acetate (pH 6.0).
2. The RNA is heat-denatured at 100°C for 2 min and snap-cooled in ice.
3. Centrifugation for a few seconds collects the reaction mixture at the bottom of the microcentrifuge tube.
4. The mRNA is then decapped by the addition of 1 μl 100 mM DTT and 1 μl (2.9 U) tobacco acid pyrophosphatase and incubated at 37°C for 30 min.
5. Dephosphorylation of the decapped mRNA is achieved by the addition of 2 μl 0.5 M Tris-HCl (pH 8.0) and 1 μl (20 mU) CAIP (Boehringer Mannheim), incubating at 50°C for 30 min.
6. The reaction is terminated by addition of 1 μl 0.25 M potassium phosphate (pH 9.5).

Once the RNA samples have been decapped they should either be labeled immediately by phosphorylation or stored at −20°C.

3′-End labeling

Methods for the 3′-end labeling of DNA. DNA can be 3′-end labeled using the enzyme terminal deoxynucleotidyl transferase (terminal transferase) which catalyzes the repetitive addition of mononucleotides using dNTP substrates to the 3′-terminus of a DNA initiator. This addition can produce a labeled tail or a single incorporated nucleotide when chain terminators are used (e.g. dideoxynucleosides such as [α-^{32}P]cordycepin-5′-triphosphate).

Protocol 3.15: Labeling by the addition of [α-^{32}P]dNTP tails to 3′-termini of a single-stranded DNA primer

1. Labeling reaction:
   4 μl 5 × terminal transferase buffer (500 mM cacodylate (pH 6.8), 1 mM $CoCl_2$, 0.5 mM DTT, 500 μg/ml BSA)
   2 pmol primer termini
   1.6 μl [α-^{32}P]dATP (29.6 TBq (800 Ci), 370 MBq (10 mCi)/ml)*
   1 μl terminal transferase
   Adjust the volume to 20 μl using sterile double-distilled water.
2. Labeling is carried out by incubation at 37°C for 30–60 min. The reaction is terminated by heating to 70°C for 10 min.

*The ratio of concentration of primer to label determines the length of tail added to the primer. For example 10 pmol labeled nucleotide used in a reaction with 1 pmol primer will produce tails 10 nucleotides in length with a specific activity greater than 1×10^9 c.p.m./μg.

This method may also be applied to double-stranded DNA, preferably with a protruding 3′ terminus. DNA fragments possessing recessed 3′ termini or blunt-ended molecules will not be uniformly labeled using this method.

Protocol 3.16: Labeling by the addition of [α-^{32}P]cordycepin-5′-triphosphate to 3′-termini of a single-stranded primer

1. Labeling reaction:
   10 μl 5 × terminal transferase buffer
   10 pmol termini
   7.5 μl [α-^{32}P]cordycepin-5'-triphosphate (111 TBq (3000 Ci), 370 MBq (10 mCi)/ml)
   2 μl terminal transferase
   To a final volume of 50 μl using sterile, double-distilled water.
2. Labeling at 37°C for 10 min precedes reaction termination by heating to 70°C for 10 min.

Methods for the 3′-end labeling of RNA. The 3′ terminus of tRNA can be labeled using poly(A) polymerase to add labeled cordycepin triphosphate. The protocol used follows.

Protocol 3.17: Labeling by incorporation of cordycepin-5′-triphosphate using poly(A) polymerase

1. The reaction is set up with 100 pmol free 3′ ends. Native tRNA molecules are highly folded and this inhibits labeling. Therefore, heat denaturation is used to unfold the molecules by boiling. To denature the RNA mix 5 μl RNA solution with 1 μl 100 mM Tris-HCl, boil for 2 min and snap-cool in ice.
2. Labeling is initiated by transferring the heat-denatured tRNA to a microcentrifuge tube containing 1.85 GBq predried 5′[α-^{32}P]-cordycepin-5′-triphosphate and adding 1 μl 670 mM $MgSO_4$, 1 μl 20 mM $MnCl_2$, 2 μl (4 U) poly(A) polymerase. This reaction mixture is incubated at 37°C for 60 min to label the 3′ end of the RNA.
3. The reaction is terminated by ethanol precipitation (10 μl 4 M ammonium acetate (pH 7.0) and 50 μl ethanol at −20°C overnight).
4. Centrifugation at 12 000 *g* for 10 min pellets the RNA. Unincorporated 5′[α-^{32}P]cordycepin-5′-triphosphate remains in the supernatant which is discarded. Any remaining unincorporated nucleotide can be removed by passing the RNA over a spin-column (Section 3.5.1, Nick translation).

3.5.2 Blotting procedures for analyzing gels

The transfer of RNA from a gel on to a membrane is called Northern blotting. In this procedure most of the RNA in the gel is transferred to a membrane either by capillary blotting alone or by accelerated blotting (e.g. electroblotting and vacuum blotting) *Figure 3.11*. Once transferred and fixed to the membrane the position and amount of the RNA present can be determined either directly if the RNA is radioactive or by use of a labeled probe to a specific sequence. The type of membrane used depends on the resolution needed and the number of sequential hybridizations (if any) to be carried out. Nitrocellulose and nylon membranes are the two commonest types of membrane used in RNA blotting. Nitrocellulose membranes generally give better resolution but are rather brittle, necessitating extreme care whilst handling and

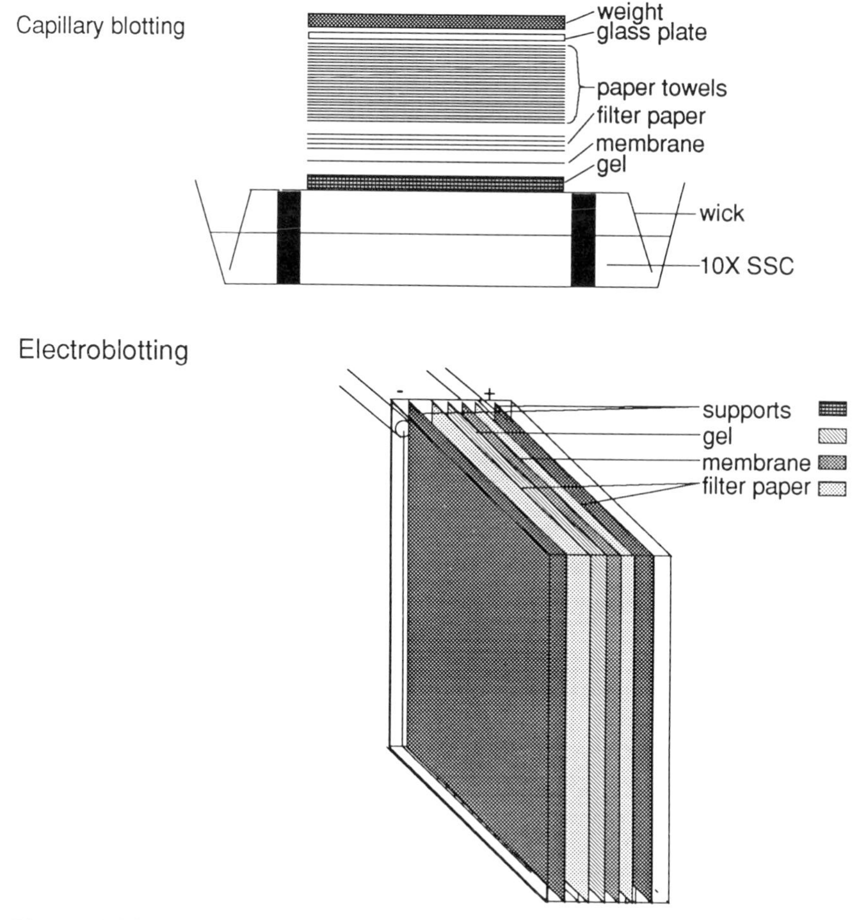

Figure 3.11: Transfer of RNA to membranes using capillary or electroblotting (Northern blotting).

restricting their use for sequential hybridizations. Nylon membranes are the routine choice for analytical hybridizations as their durability and low cost favor their use. *Table 3.18* lists some commercially available nylon membranes suitable for RNA analysis.

Table 3.18: A selection of commercially available nylon membranes used for blotting gels

Commercial name	Supplier
Hybond-N^{+}	Amersham
Zeta-probe	Bio-Rad
Zeta-probe GT	Bio-Rad
Genescreen	Du Pont–NEN
Genescreen Plus	Du Pont–NEN
Optiblot	IBI
Biodyne A	Pall
Nytran	Schleicher and Schuell
Duralon-UV	Stratagene

After transfer, the RNA species must be fixed to the membrane to reflect their positions in the gel. The two main methods of accomplishing this are baking and UV cross-linking. Both methods fix the nucleic acid backbone to the membrane matrix. Baking the membrane at 80°C for 2–4 h fixes the RNA to the membrane whilst UV cross-linking takes 15 sec to 2 min. Although baking is slower it is generally more reliable as UV fixation can lead to high backgrounds if membranes are exposed to UV for long periods of time. Automatic exposure of membranes can be carried out where fluctuations in UV strength are measured and exposure time is adjusted. This can be achieved by a machine such as the UV Stratalinker 1800 (Stratagene) which cross-links the RNA with the membrane in 15–30 sec. UV cross-linking performed in this manner can give better fixation than baking in an oven. Once the RNA is fixed to the membrane, hybridization to radiolabeled molecules (see Section 3.5.1) or to chemiluminescent systems (Section 3.5.3) is possible.

Hybridization conditions. Hybridization using a labeled probe can be used to identify the amount and location of specific RNA sequences. The process which underlies all of the methods based on molecular hybridization of nucleic acids is duplex formation of two complementary strands. The factors that affect the rate of hybrid formation are described in Section 4.3.2. However, RNA molecules are more susceptible to degradation, particularly by high temperatures and in alkaline solutions. RNA hybridizations are therefore usually carried out at lower temperatures than standard DNA hybridizations and use hybridization buffers where the pH is less than 8. Formamide-containing buffers are standardly used to preserve RNA integrity (*Table 3.19*).

Table 3.19: Formamide-based hybridization buffers

Constituents
1. 50% formamide, 5 × SSPE[a], 2 × Denhardt's[b], 0.1% SDS, 0.1 mg/ml herring sperm DNA[c]
2. 50% formamide, 5 × SSPE, 5 × Denhardt's, 0.5% SDS, 0.1 mg/ml herring sperm DNA

[a]20 × SSPE stock solution: 3 M NaCl, 0.177 M NaH_2PO_4, 0.02 M EDTA adjusted to pH 7.4 with 10 M NaOH.
[b]100 × Denhardt's: 500 ml: 10 g Ficoll, 10 g polyvinylpyrrolidine, 10 g BSA.
[c]Herring sperm DNA is given here as the blocking agent to prevent non-specific binding of labeled probes to the membrane. This can be replaced with many other DNA sources such as bacterial, salmon sperm DNA and calf thymus DNA as appropriate.

The formation of duplex structure between the probe and the target sequence is very sensitive to a number of factors especially temperature. Too high a temperature will prevent the formation of any hybrid and too low a temperature will allow mismatching between the probe and target sequence. Hence it is critically important to select temperatures that allow the probe to form the closest to a duplex structure possible for it. Hybridization temperatures can be calculated using an equation based on the 'melting temperature' (T_m) of the nucleotide probe being used (Section 4.3.2). Generally different hybridization temperatures are used for homologous and heterologous probing. Homologous probing uses probes that have 100% identity to the target sequence as the target and the probe come from the same organism. The temperatures used are therefore quite stringent, normally around 42°C. Heterologous probing uses hybridization probes from a different organism to the one being probed. The matching of these to the target sequence is rarely perfect and often lower hybridization temperatures of about 37°C are used.

The volume of hybridization mixture should be kept to a minimum so as to maximize the concentration of the probe. As described in Section 4.3.2, the rates at which hybrids form are dependent on concentration and time and so a low concentration of labeled probe in the hybridization solution leads to low levels of hybridization unless a long incubation time is used. As a guide, for each 100 cm^2 of membrane use 15 ml hybridization solution for prehybridization and 10 ml for the hybridization reaction itself.

Hybridization systems for membranes must allow contact of hybridization mixture to all areas of the membrane. Two hybridization systems are used for ensuring equal spread of hybridization mixture to all areas of the membrane. The first uses polythene heat-sealable bags which can hold the membrane and hybridization solution. When using this system it is important that all the air bubbles are expelled as they can prevent areas of membrane being subject to prehybridization or hybridizing with a probe which would cause uneven exposure to the probe. The second type of system uses round hybridization chambers or bottles.

Membranes are placed in the Pyrex bottles or in the tank along with the required amount of solution. Bottles can then be placed in a rotisserie type of incubator (e.g. Hybaid minioven, *Figure 3.12*) which rotates the bottles, ensuring an even spread of the hybridization solution over the membrane. The use of Pyrex containers avoids the danger inherent in plastic bag manipulations and shields the user from radioactive emissions.

Figure 3.12: Hybridization oven.

Protocol 3.18: Prehybridization of membranes

1. Membranes are prehybridized with the hybridization solution, which has no labeled probe included, to prime the membrane for hybridization by blocking the sites on the membrane that can bind labeled probes non-specifically thus giving rise to high backgrounds in subsequent analyses.
2. If using polythene bags, the membrane and prehybridization solution are placed in the bag, air is excluded and the bag is sealed with a heat sealer, placed within another bag which in turn is heat sealed and then incubated. Incubators must ensure equal spread of mixture to all parts of the bag. When using bottles the membrane is placed around the wall and the solution is added; take care to ensure that there are no air bubbles between membrane and the wall. Orbital incubators or hybridization ovens should move the solution evenly over the membrane. It is important to prevent any leaks from the bags or bottles.

3. Prehybridization is usually performed at the hybridization temperature (e.g. 37°C or 42°C) for 1–2 h to ensure that all the sites on the membrane that could non-specifically bind the probe are blocked.

Protocol 3.19: Hybridization using a labeled probe

1. The prehybridization solution is discarded and fresh hybridization solution, prewarmed to the hybridization temperature, is added.
2. The radiolabeled probe (e.g. ^{32}P) is heat-denatured and snap-cooled in ice before use to allow subsequent duplex formation with target RNA. The probe should be carefully pipetted into the hybridization solution within the hybridization container (plastic bag or hybridization chamber). If the plastic bag system is being used the hybridization solution should be confined in an area equal to the size of the membrane. This can be achieved by sealing the bag around the membrane before adding the hybridization solution.
3. Hybridization is carried out for a period of 4–16 h depending on the nature of the probe, target sequence and the degree of complementarity between probe and target. For example, a high-specific-activity probe that has perfect complementarity to a target species affixed to the membrane can generate a signal after 4 h of hybridization whilst low-specific-activity probes or those showing less than perfect complementarity to target species (e.g. heterologous probing) probably need to be hybridized overnight.
4. Hybridization solution must be disposed of strictly according to current legal guidelines, either down a radioactive disposal sink or bottled for disposal. For sink disposal copious amounts of water should be used to flush away radioactivity from the sink area. Care should be taken to monitor disposal areas for radiation before and after use and decontaminate any radioactive areas as necessary.
5. All glassware should be decontaminated by soaking in a concentrated solution of laboratory detergent. If using plastic bags, these must be thoroughly rinsed after the experiment, to wash away any remaining radioactive hybridization solution before they are disposed of. The plastic bag system is inexpensive yet manipulations are slightly awkward and so there is the potential danger of spillage or leakage.
6. Unbound and non-specific radioactivity is washed off the membrane with a series of washes of increasing stringency. Stringency is increased by washing with solutions of lower salt concentration and/or by use of higher temperatures. Examples of some typical washing procedures used are detailed in *Table 3.20*.

Table 3.20: Selected washing solutions and conditions for removal of non-specifically bound radiolabeled probes

	Washing solution	Conditions
1.	a. 1 × SSC, 0.1% SDS	Room temperature/20 min/wash once
	b. 0.2 × SSC, 0.1% SDS	68°C/10 min/three washes
2.	a. 1 × SSC, 1% SDS	Room temperature/10 min/wash twice
	b. 0.2 × SSC, 0.1% SDS	65°C/10 min/wash twice

SSC is 0.15 M NaCl, 0.015 M sodium citrate pH 7.0.

Protocol 3.20: Autoradiography

1. The membrane is placed on a support sheet (usually a piece of used X-ray film covered with cling film) and wrapped with cling film to ensure that it remains moist. The reason for this is that if the membranes dry out then it is not possible to strip off the hybridized radioactive probes to allow subsequent analyses.
2. Membranes are then exposed to X-ray film in a light-tight cassette at −70°C (^{32}P) or room temperature (^{35}S); for the former the use of intensifier screens will enhance sensitivity. For further details on autoradiography and fluorography see later.
3. Developing of the X-ray film may be performed manually, transferring the film from developer to fixer and rinsing, or by use of an automatic developer. The exact protocol to be used should follow the manufacturers' instructions.

Some membranes can be stripped after analysis allowing them to be reprobed for other RNA sequences. This procedure involves pouring a stripping solution heated to boiling point over the membranes and rocking or shaking them for 10 min. Various stripping solutions are used, including 0.1% SDS or 1% SDS in Tris-EDTA buffer (10 mM Tris-HCl, pH 7.5, 1 mM EDTA).

Choice of the type and labeling of probes for use in Northern hybridization. Nucleic acid probes for Northern hybridization may be RNA or DNA. RNA probes form hybrids with the bound RNA which are more stable than DNA:RNA hybrids. The need to perform sequential probings often favors DNA probes as they are easier to strip from the membrane after probing. Probes for use on Northern blots are labeled by a variety of methods described in Section 3.5.1 and summarized in *Table 3.21*.

Autoradiography and fluorography. There are three methods of autoradiographic assessment. Direct autoradiography involves the exposure of the radioactive gel or membrane blot to X-ray film, producing band positions corresponding to the position of the

Table 3.21: Labeling methods for use in Northern hybridization

Method basis	RNA/DNA probe	Comment
Nick translation	DNA	Used to generate labeled DNA probes of moderate specific activity
RNA polymerase	RNA	Often used to generate RNA transcripts from plasmids. RNA polymerase sites flanking plasmid inserts can be used to synthesize sense or antisense probes. RNA:RNA hybrids are quite stable and are of use in detecting low-level species or in heterologous probings
DNA polymerase	DNA	Commonly used for preparing probes for Northern blots. The standard method is based around the random hexanucleotide primed method [16] generating many different sizes of fragment complementary to the target RNA
End labeling	DNA/RNA	Used when oligonucleotide probes are to be used. An oligonucleotide of 25–50 nucleotides is commonly 5′-end labeled by T4 polynucleotide kinase but 3′-end labeled probes can also be prepared
In vivo	RNA	RNA synthesis over time can be measured by incubating nuclei with radioactive nucleotide precursors before hybridizing to known concentrations of RNA on a Northern blot

radioactivity. Fluorography uses scintillants to increase the sensitivity of the film for weak radioisotopes such as tritium. Indirect autoradiography causes radiation not absorbed by X-ray film to be converted to light, which in turn has effect on the X-ray film.

The method of assessment depends on the nature of the isotope being used (*Table 3.22*). Radioactive isotopes such as ^{32}P and ^{125}I have emissions strong enough to penetrate the gel segment in which they are present. These isotopes emit particles of such strength that few are absorbed by X-ray film but instead pass straight through the film, thus generating an image that is not representative of the amount of radiation present. In these cases, indirect autoradiography should be

Table 3.22: Isotopes used for labeling experiments

Isotope	Half-life	Emission type	Energy of emission (MeV)		Typical sensitivity[a] (d.p.m./cm^2)	
			per particle	per γ-ray	Auto-radiography	Indirect auto-radiography
^{3}H	12.3 years	β	0.0185	–	–	–
^{32}P	14.3 days	β	1.71	–	500	50
^{33}P	25.4 days	β	0.249	–	1250	–
^{35}S	87.1 days	β	0.169	–	6000	–
^{125}I	60 days	γ	–	0.035	1600	100

[a]Necessary to produce an image on X-ray film upon 24 h exposure.

used to locate the radioactive bands. This can be achieved by placing an 'intensifying' screen behind the X-ray film. Intensifying screens incorporate calcium tungstate which emits multiple flashes of light when radioactive particles impact on it, thus producing a photographic image overlaying the autoradiographic image. Note that the autoradiographic images of isotopes which emit very penetrating radiation have a poorer resolution than those of weaker isotopes. For example, images from ^{35}S-labeled probes are sharper than those of ^{32}P-labeled probes.

Isotopes with weaker emissions such as ^{3}H, ^{33}P, ^{35}S and ^{14}C cannot penetrate the gel matrix requiring the gel to be dried before autoradiographic assessment. Gels are usually dried by using commercial gel driers with heating elements. This type of apparatus can dry gels to a thin film in only 1–2 h. Gel driers that use vacuum systems only, can take up to 24 h to dry gels. Labeled species containing ^{3}H in gels cannot be detected by autoradiography even if the gel is dried. In order to detect weak isotopes it is necessary to use fluorography; this method can also be used to enhance the detection of other isotopes such as ^{14}C and ^{35}S. For fluorography the gel is infiltrated with a scintillant which emits light from within the gel and thus exposes the X-ray film. Some methods are outlined in *Table 3.23* and some commercially available products such as Enhance (NEN) are also available.

Gels are normally exposed to X-ray film for a period of time according to their radioactivity. In the case of ^{32}P-labeled gels, a Geiger–Müller tube monitor can be used as a rough measure of the radioactivity contained in the gel before autoradiography. As a guide, radioactivity levels of 100–200 c.p.s. of ^{32}P need only be exposed to X-ray film for about an hour whilst low levels of between 5 and 20 c.p.s. can need as much as 2 days or more exposure.

When gel segments are to be measured for their radioactive content, the nature of the isotope must again be taken into consideration. Segments containing a strong emitter such as ^{32}P or ^{125}I can be measured in a scintillation counter without the aid of scintillants. This method of counting utilizes the beta emissions of ^{32}P which cause Cerenkov radiation. This radiation is usually measured using the ^{3}H channel on a scintillation counter. The efficiency of counting is at most 50% and varies with gel segment size. Segment size distortions can be bypassed by placing the gel segment in a normal scintillation vial with 0.5 ml 1 M NaCl before counting [15]. Segments containing low-level radiation can be dried down on filter paper and measured against radioactive standards or the RNA can be eluted (Section 3.6), mixed with scintillants and counted in a scintillation counter.

Table 3.23: Fluorographic detection of radioactivity in gels

Scintillant and gel type	Procedure
PPO[a] with polyacrylamide gels	A. Immerse stained or unstained polyacrylamide gel in 20 times its volume of dimethyl sulfoxide for 30 min; repeat in fresh solution. B. Immerse gel in 4 vol. 20% PPO (w/w) dissolved in dimethyl sulfoxide for 3 h. C. Wash gel in 20 vol. water for at least 1 h to remove dimethyl sulfoxide and precipitate PPO in the gel. D. Dry the gel under vacuum. E. Place the dried gel in contact with preflashed Kodak X-omat R film at −70°C. Caution: Dimethyl sulfoxide is toxic, avoid inhalation and contact with skin.
PPO with agarose gels or composite gels (<2% agarose)	A. Immerse stained or unstained gel in 20 times its volume of 100% methanol for 30 min; repeat in fresh solution. B. Immerse gel in 4 vol. 20% PPO (w/w) dissolved in 100% methanol for 3 h. C. Wash gel in 20 vol. water for at least 1 h to remove 100% methanol and precipitate PPO in the gel. D. Dry the gel under vacuum. E. Place the dried gel in contact with preflashed Kodak X-omat R film at −70°C.
Sodium salicylate	A. Fix the gel if required, in 5% TCA or methanol:acetic acid:water (5:1:5 by vol.). B. If the gel has been fixed, the acid must be removed by immersing the gel in 20 vol. distilled water for 30 min; this prevents sodium salicylate precipitation. C. Immerse the gel in 10 vol. 1.0 M sodium salicylate (pH 7.0) for 20 min. D. Lay the gel on moist Whatman 3MM paper and dry under vacuum. E. Proceed with autoradiography.

[a]PPO, 2,5-diphenyloxazole.

Quantitation of autoradiographs. Measurement of the amount of labeled RNA can be achieved by correct use of densitometry equipment. The sensitivity of densitometers is, however, sometimes lower than expected, even in the case of laser densitometers. Bands visible to the eye are sometimes not recognized as being above background and, as such, are not registered as bands. Autoradiographs are scanned over a defined area of the densitometer's grid system. Data is transferred to a computer software package which generally offers many useful options (*Table 3.24*). A fact of primary importance is definition of background density. This varies with exposure time, X-ray film type, film age and background radiation on the sample support (gel or membrane). Software packages usually can designate molecular sizes to bands by use of the mobility of standards.

Computer assessment should, when possible, be checked by returning to the autoradiographs and getting experienced assessment by eye as to RNA size. Measurement of a couple of autoradiographs of the same samples but at different exposure times allows a double-check of the defined size.

Table 3.24: Software systems for use in gel imaging

Package name	Supplier	Comments
ImageQuant™ v3.3	Molecular Dynamics	Scanning control, volumetric density integration, windows compatibility, IBM or Macintosh versions (DOS system)
Kepler	Large Scale Biology Corporation	Scanning control, spot extraction, quantitation, database comparison (2D patterns), VAX-based system for autoradiographs derived from labeled protein or DNA
PDQUEST	Protein and DNA Imageware Systems	Scanning control, spot extraction, quantitation, database comparison (2D patterns), UNIX-based system for autoradiographs derived from labeled protein or DNA

3.5.3 Chemiluminescent localization of RNA on membranes

Chemiluminescence is becoming a favored method of localization and assessment of nucleic acids. The procedure centers on conjugation of biotin to the protein avidin. Avidin can then be used as an epitope site for anti-avidin antibodies. These antibodies can either have reporter enzymes such as horse radish peroxidase (HRP) or alkaline phosphatase attached or can form the epitope for a secondary antibody which has a reporter enzyme present. Biotin can be conjugated to nucleic acids by a method called biotinylation. Biotin may be attached to nucleic acids enzymatically by incorporation of biotin-labeled nucleotides (e.g. biotinylated dUTP) or chemically (*N*-hydroxysuccinimide–long chain biotin $(CH_2)_9$ with an oligonucleotide) to preformed oligonucleotides by photobiotinylation (Section 2.9). Hybridization of biotinylated nucleic acid specific to the RNA of interest on a Northern blot bearing separated RNA allows localization of RNA bands by staining or chemiluminescence. Staining methods can only be performed for one RNA species as the stains cannot be removed. *Table 3.25* lists staining methods used for the detection of biotin-labeled RNA.

A commercial system which uses HRP as a reporter enzyme is called enhanced chemiluminescence (ECL) and can be purchased from Amersham International. This system in the presence of hydrogen peroxide can oxidize the substrate luminol (*Figure 3.13*). Luminol is

Table 3.25: Some staining procedures using reporter enzymes

Reporter enzyme	Substrate	RNA band colour
Alkaline phosphatase	5-Bromo-4-chloro indoxyl phosphate (BCIP) and nitroblue tetrazolium	Purple
Horseradish peroxidase	4-Chloro-1-naphthol	Purple
Horseradish peroxidase	Diaminobenzidine (DAB)	Brown

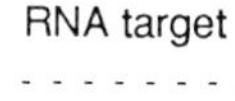

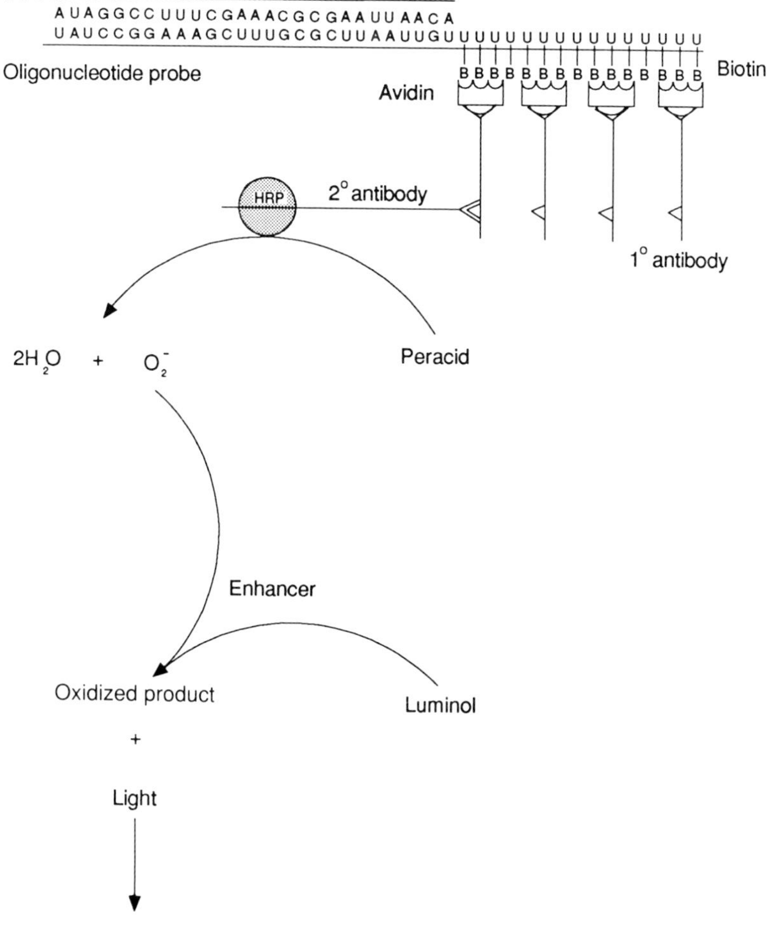

Figure 3.13: ECL detection of specific RNA species.

raised to an excited state by this reaction and returns to its ground state with the emission of light. The enhancement integral to the system is accomplished by a phenolic enhancer which increases the intensity of emitted light by up to 1000 times. The reaction peaks after 1–5 min and then declines. A permanent record can be produced by covering the membrane with Saran Wrap (to keep any moisture away from the X-ray film), placing the covered membrane in contact with a piece of X-ray film for a few seconds (exposure time varies with emission intensity) and developing immediately. Various exposures are carried out at the one time to ensure that a correct exposure is achieved.

The other main reporter gene, alkaline phosphatase can be used to cleave substrates with a concomitant emission of light. The two substrates commonly used are disodium 3-(4-methoxyspiro[1,2-dioxetane-3,2′-tricyclo-[3.3.1.$1^{3,7}$]decan]4-yl)phenyl phosphate (AMPPD) and a 5-chloro derivative (CSPD) [19]. The reporter enzyme, alkaline phosphatase, is linked to streptavidin which can bind to biotin. Biotinylated oligonucleotide probes can bind to streptavidin–alkaline phosphatase conjugates to form an enzymatically active complex in a very similar manner to the ECL method described previously. The phosphatase/dioxetane reaction involves dephosphorylation of the substrate producing the destabilized dioxetane anion which quickly decomposes to the chemiluminescence-generating compound methyl *m*-oxybenzoate. As methyl *m*-oxybenzoate undergoes transition to the ground state, light is emmitted with a maximum at 470 nm. Enhancers for the emission are also available for this system. Detection and storage of intensity can be accomplished using X-ray film or by scanning with a light-sensitive charge-coupled (CCD) camera.

3.6 Recovery of RNA from gels

The yield of RNA from electrophoretic gels is not as high as that which can be achieved using preparative centrifugation but it is often easier to use electrophoretic separation. The two electrophoretic gel matrices, polyacrylamide and agarose are treated differently when RNA recovery is required. Two types of elution method can be applied to these gels. Electroelution causes the RNA to move out of the gel segment that contains it along an electrical field, into the surrounding buffer which is then recovered. Chemical elution involves solubilizing the gel in elution buffer and incubating for a period of time. The RNA can then be recovered.

3.6.1 Recovery of RNA from polyacrylamide gels

RNA is usually recovered from polyacrylamide gels by simple diffusion. The elution of RNA by diffusion can be accomplished by various buffers, some of which are listed in *Table 3.26*. Fragments are usually eluted into

Table 3.26: Elution buffers used for polyacrylamide gel segments

Elution buffer	Comment
0.15 M sodium acetate (pH 6.0) and 0.5% SDS	
0.6 M lithium acetate (pH 6.0) and 0.5% SDS	Frequently used for tube gel segments
0.8 M lithium acetate (pH 6.0) and 0.5% SDS	
2 × SSC with or without 0.2% SDS	Use of SSC buffers allows immediate use of the extracted RNA in subsequent hybridization studies
6 × SSC with or without 0.2% SDS	

SSC is 0.15 M NaCl, 15 mM sodium citrate pH 7.0.

as small a volume as possible with a guideline of 140 mm^3 fragments (tube gel of 6 mm diameter and 5 mm depth or slab gel of 2 mm depth, 7 mm height and width) being eluted into 0.75 ml buffer. Fragments can be kept whole or homogenized in the elution buffer. Elution is normally carried out at 30°C for at least 5 h.

The extracted RNA can be purified from gel segments by extraction with an equal volume of phenol/cresol. The phases are mixed and centrifuged for 10 min at 10 000 *g*. The RNA in the recovered supernatant is then precipitated by the addition of a one tenth volume of 3 M sodium acetate (pH 5.2) and 2 vol. ethanol. RNA can also be recovered from gels by electroelution.

Protocol 3.21: Electroelution of RNA from gel segments

1. The gel segment is placed into a length of visking tubing and an equal volume of electrophoresis buffer is added.
2. The visking tubing is sealed with clips and placed into the electrophoresis tank with the gel segment closest to the anode. This will allow the RNA to migrate towards the cathode into the buffer in the tubing.
3. Electroelution is usually carried out at 80 mA/90 V for 30 min.
4. A 20–30 sec reversal of the polarity of the electrical field at the end of elution recovers RNA that has adhered to the visking tubing.

3.6.2 Recovery of RNA from agarose gels

RNA can be recovered from agarose gels by simple diffusion or electroelution using exactly the same procedures as are used for polyacrylamide gels. However, the major advantage of agarose gels is that they can be liquified simply by moderate heating. Hence, recovery of RNA from agarose gels is normally accomplished by use of low gelling temperature agarose (LGT). This type of agarose will melt at a reasonably low temperature around 70°C and remain liquid at 37°C. Gels containing urea will remain liquid at 20°C which is more convenient for the following extraction procedures. Extraction is based around the use of hexadecyl-trimethyl ammonium bromide (HTAB).

Protocol 3.22: Elution of RNA from agarose gels

1. The molten gel is extracted three times with equal volumes of 3.67% (w/v) HTAB in 1-butanol.
2. The extracts are combined and 1/4 vol. 0.2 M NaCl is added.

Lower aqueous phase is taken and the HTAB–1–butanol is re-extracted with 1/4 vol. 0.2 M NaCl.
3. An equal volume of chloroform is gradually added to the collected aqueous solution and the HTAB is precipitated by placing on ice.
4. RNA can be precipitated with ethanol and redissolved in desired solution.

Gels containing methyl mercuric hydroxide need to be treated in a different way.

Protocol 3.23: Elution of RNA from gels containing methyl mercuric hydroxide

1. Gels are soaked in 0.1 M DTT for 30–40 min.
2. The gel is cut into 3 mm segments and each is inserted into separate microcentrifuge tubes.
3. Gel segments are dissolved in 4 vol. 0.5 M ammonium acetate, preheated to 65°C and maintained at 65°C until fully dissolved.
4. Extraction with an equal volume of phenol (pH 7.6) and centrifugation at 2000 *g* for 10 min in a cold-room precipitates the agarose at the interface.
5. Two chloroform extractions remove any residual agarose.
6. Agarose can be precipitated with 0.1 vol. 3 M sodium acetate (pH 5.2) and 3 vol. ethanol.

3.7 Use of sequencing to determine the size of RNA

3.7.1 Sequencing genes

The only definitive method of assessing RNA size is to sequence the RNA of interest. The extensive developments in rapid DNA sequencing methods have made it possible to characterize large numbers of gene sequences and, given that it is possible to recognize important motifs such as the initiation of transcription and intron/exon splice sites, it is theoretically possible to predict quite accurately the sequence, and hence the size, of mRNA derived from each gene. In fact, the situation is much more complicated not only because of the difficulties in recognizing the precise initiation and termination sites used by the RNA polymerases, but also because mRNA can undergo a wide range of post-transcriptional processing events such as polyadenylation and editing of the sequence which are difficult to predict from the gene sequences. It is for this reason that it is better to analyze the sequence of the RNA itself by direct sequencing methods.

3.7.2 Sequencing RNA

Original estimates of RNA size were based on the alkaline hydrolysis of labeled RNA as a result of which the 5′-terminal nucleotide should have multiple phosphate groups and the 3′-terminal nucleotide none, while all internal nucleotides have a single phosphate group. Hence, by separating the hydrolysis products and calculating the ratio of internal to terminal nucleotides, it is possible to obtain an estimate of the size of the RNA. However, the advent of fast, automated sequencing methods for RNA now allows the researcher to analyze the size of RNA molecules with complete precision. RNA sequencing is described in detail in Chapter 4.

References

1. **Spragg, S.P. and Steensgaard, J.** (1992) in *Preparative Centrifugation: a Practical Approach* (D. Rickwood, ed.). IRL Press, Oxford. p. 1.
2. **Rickwood, D. (ed.)** (1992) *Preparative Centrifugation: a Practical Approach.* IRL Press, Oxford.
3. **Rickwood, D.** (1992) in *Preparative Centrifugation: a Practical Approach.* (D. Rickwood, ed.). IRL Press, Oxford. p. 143.
4. **Sambrook, J., Fritsch, E.F. and Maniatis, T.** (1989) *Molecular Cloning—A Laboratory Manual,* 2nd edn. Cold Spring Harbor Laboratory Press, New York.
5. **Grierson, D.** (1990) in *Gel Electrophoresis of Nucleic Acids: a Practical Approach,* 2nd edn. (D. Rickwood and B.D. Hames, eds). IRL Press, Oxford. p. 1.
6. **Peacock, A.C. and Dingman, C.W.** (1968) *Biochemistry,* **7**, 668.
7. **Dahlberg, A.E. and Grabowski, P.J.** (1990) in *Gel Electrophoresis of Nucleic Acids: a Practical Approach,* 2nd edn. (D. Rickwood and B.D. Hames, eds). IRL Press, Oxford. p. 275.
8. **De Wachter, R., Maniloff, J. and Fiers, W.** (1990) in *Gel Electrophoresis of Nucleic Acids: a Practical Approach,* 2nd edn. (D. Rickwood and B.D. Hames, eds). IRL Press, Oxford. p. 151.
9. **Ikemura, T. and Dahlberg, J.E.** (1973) *J. Biol. Chem.,* **248**, 5024.
10. **Fradin, A., Gruhl, H. and Feldman, H.** (1975) *FEBS Lett.* **50**, 185.
11. **Burckhardt, J. and Birstiel, M.L.** (1978) *J. Mol. Biol.,* **118**, 61.
12. **Vigne, R. and Jordan, B.R.** (1971) *Biochemie,* **53**, 981.
13. **Stein, M. and Varricchio, F.** (1974) *Anal. Biochem.,* **61**, 112.
14. **De Wachter, R. and Fiers, W.** (1972) *Anal. Biochem.,* **49**, 184.
15. **Billeter, M.A., Parsons, J.T. and Coffin, J.M.** (1974) *Proc. Natl Acad. Sci. USA,* **71**, 3560.
16. **Feinberg, A.P. and Vogelstein, B.** (1983) *Anal. Biochem.,* **132**, 6.
17. **Taylor, G.R.** (1991) in *PCR: a Practical Approach* (M.J. McPherson, P. Quirke and G.R. Taylor, eds). IRL Press, Oxford. p. 1.
18. **Stahl, D.A., Krupp, G. and Stackebrandt, E.** (1989) in *Nucleic Acids Sequencing: A Practical Approach* (C.J. Howe and E.S. Ward, eds). IRL Press, Oxford. p. 137.
19. **Beck, S.** (1993) *Meth. Enzymol.,* **216**, 143.

4 Determination of RNA sequences

4.1 An overview of methods for determining RNA sequences

The determination of RNA sequences can be carried out in different ways depending on the type of information required. Hybridization is extremely important in that it can be used to identify the presence or absence of specific gene sequences. As the availability of diverse specific gene and oligonucleotide probes has increased so has the importance of preliminary analysis of sequences by hybridization. RNA fingerprinting using specific nucleases to produce defined fragments has also retained a measure of popularity as a way of comparing sequences without having to do a complete sequence analysis. However, it has become clear that sequencing RNA molecules is the only way to carry out analyses on the structure and function of RNA. Moreover, the discovery of RNA editing (see Section 1.6.4) has emphasized the need to determine the RNA sequence itself and not just the DNA sequence from which it was derived. RNA sequencing should now be considered to be the method of choice for the characterization of RNA.

4.2 Methods for labeling nucleic acids

The labeling of nucleic acids is an integral part of RNA analyses. Labeling methods used to rely totally upon radioisotopic methods (Section 3.5.1) but subsequently, non-isotopic labeling methods have become a realistic alternative. Most of the non-isotopic labeling methods involved the synthesis of biotinylated probes (Section 2.9.2 and Section 3.5.3). See Chapters 2 and 3 for details of the methods used for labeling DNA and RNA.

4.3 Analysis of RNA sequences using hybridization

Nucleic acid hybridization can be used to determine primary structure, abundance and rate of synthesis of an individual RNA species in a complex population as well as giving an indication of the degree of homology between different genes and organisms. The process which underlies all of the methods based on molecular hybridization is the formation of the double helix from two complementary strands. Many of the procedures used are similar to those for DNA. However, RNA molecules are much more susceptible to degradation than DNA. The major danger is that of cleavage by ribonucleases during extraction and purification. Therefore procedures to minimize this problem should always be followed (see Section 2.1). In addition, at high temperature, and especially at neutral or alkaline pH, RNA is degraded much more rapidly than DNA. Hence hybridization should be performed for the minimum length of time possible in a neutral buffer and at the lowest temperature allowing efficient hybridization. Formamide-containing buffers are a necessity rather than an option for those hybridization procedures where the integrity of the RNA is of paramount importance.

There are two basic systems for hybridization reactions. Either one of the two nucleic acid molecules in the hybridization reaction is immobilized on a solid support (membrane hybridization), or the hybridization is carried out with both nucleic acids in solution. In solution hybridization, both the target and probe nucleic acids are free to move, thus maximizing the chance that complementary sequences will align and bind. Consequently, solution hybridization reactions go to completion 5- to 10-fold faster than those on solid supports [1]. This can be particularly important in many diagnostic microbiology applications where the concentration of the target sequence is very low and speed is essential. Membrane hybridization systems are more commonly used due to their ease of handling and analysis compared with solution hybridization methods and so more emphasis will be placed on the applications of membrane hybridization.

4.3.1 Probes for hybridization analyses

The probes can be radiolabeled (Section 3.5.1) or labeled non-isotopically using biotinylation (Sections 2.9.2 and 3.5.3). A probe may be either single-stranded DNA or RNA. There are advantages to be gained from using an RNA probe as RNA:RNA hybrids are more stable than DNA:RNA and DNA:DNA hybrids. The choice of probe used depends on:

(i) target RNA abundance;
(ii) availability of information on target species;
(iii) sensitivity required;
(iv) requirements for multiple probings;
(v) safety aspects in probe production.

These factors are discussed next and summarized in *Figure 4.1.*

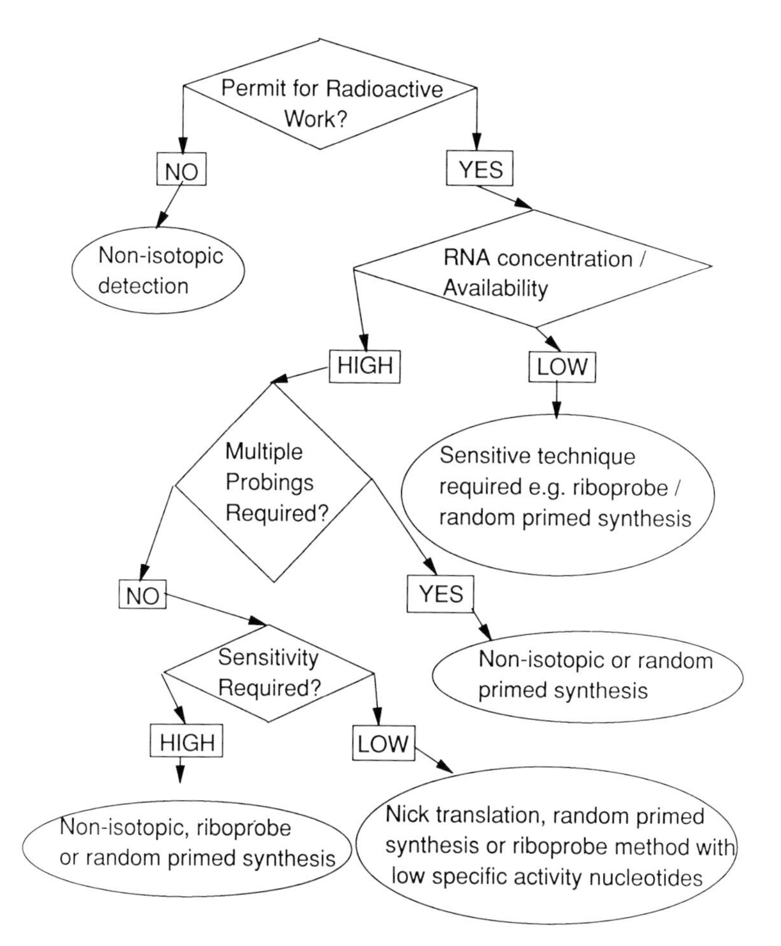

Figure 4.1: Choice of probe labeling for hybridization studies.

Target RNA abundance. The abundance of the RNA species under study to a large extent defines which probes can be used in hybridization studies. Highly abundant RNA species can be studied using less sensitive techniques as well as highly sensitive methods (if desirable). *Table 4.1* details the typical minimum amounts of RNA detectable by each of the three most popular methodologies. Rarer RNA species must be studied with more sensitive techniques in order to detect the formation of hybrids.

Table 4.1: Sensitivity of selected probing systems

System basis	Sensitivity (ng/mm^2)	Isotopic/non-isotopic
Biotin–avidin–HRP	0.5	Non-isotopic
^{32}P random primed probes	0.1	Isotopic
Riboprobes	0.05	Isotopic

Availability of information on target species. Information on the target species sequence determines the range of probes that can be used. If complete sequence identity of the RNA species is known and complementary DNA templates are available then there is no restriction on the type of probe that can be used. For example, if a certain mRNA species is under study and the cDNA is available, then preferred methods would include the random priming method (Section 3.5.1), a radiolabeled RNA probe generated from a transcription site in the vector containing the cDNA (Section 3.5.1) or a biotinylated DNA fragment or oligonucleotide (Section 2.9.2). If, however, sequence information is incomplete then only certain techniques can be used. For example, if sequence information on an RNA species of interest is only partial and encompasses only 100 nucleotides of information, then random priming is precluded as small templates yield probes that often give a high background on the membrane.

Sensitivity required. Requirements of experiments can include detection and/or quantitation of one or a series of RNA species. If detection of an RNA species is the only requirement, then the sensitivity of the probe used is only defined by the limit imposed by the abundance of the target RNA. For example, detection of a highly abundant RNA species in a sample can be accomplished by any of the procedures listed in *Table 4.1*. However, a low sensitivity method would be favored in order to reduce the manipulations involved in the hybridization. A more sensitive method is required to detect an RNA species that is in low abundance. If detection is not the only requirement of the experiment, but quantitation is the emphasis, then a highly sensitive method is required to ensure that the signal produced at the end of the hybridization procedure can be related directly to the abundance of RNA.

Requirements for multiple probings of RNA sequences. Hybridization involving multiple probings of RNA on membranes is usually accomplished by using DNA probes, and it is important to use probes that can easily be removed, allowing further probings without leaving a confusing background of the old signal. Radioactive probes (e.g. random primed synthesis) or non-isotopic probes can both be used in multiple probing protocols and, as will be described, the latter probes that utilize chemiluminescence do have some advantages over radioactive probes. If sample RNA is plentiful then multiple probing of the same samples is often not required as aliquots may be probed in isolation with different probes.

Probes can be removed from a membrane after autoradiography by 'stripping' the probe from the membrane. Stripping the probe from the membrane is accomplished by pouring boiling 0.1% SDS over the membrane and placing the membrane in the hot SDS solution on to a rocking table for 10 min. An alternative stripping procedure, that is thought to remove less of the fixed RNA from the membrane, involves placing bolts in 1 mM Tris (pH 8), 1 mM EDTA and 0.1 × Denhardt's at 75°C for 2 h. One or two strippings are usually necessary for the complete removal of a probe. Stripping also can remove some RNA from the membrane reducing the target molecule number for future probings. Thus, probes that are easier to strip from a membrane (small probes and those that hybridize to low copy number targets) should be the initial probes and probes producing large signal responses should be used at the end of the series of hybridizations. A check on the stripping procedure should be carried out each time by exposing the stripped membrane to X-ray film for the same exposure time as will be used for the next probe of the series. If no signal is produced on the film then any signal after the next hybridization must be the result of the new probe.

Non-isotopic probings have the advantage that they do not have to be stripped from the membrane. Chemiluminescent signals (e.g. derived from the dioxethane ion destabilization or luminol hydrolysis) are short lived (maximum 30 min) and further probings can be carried out as soon as the label has been exhausted (Section 3.5.3).

Multiple probings of RNA in solution by RNA probes (riboprobes) require multiple samples as the riboprobe method (Protocol 4.4) digests RNA that is not bound to a probe. Riboprobe methods therefore require a relatively large amount of RNA for multiple probings.

Safety aspects of preparing labeled probes. Because of the inherent risks in using radioisotopes, the use of radioactive materials is regulated in most countries. It is important that the reader is aware of the controls on the use of radioisotopes and does not contravene regulations. The risks of working with radioactivity can be minimized by keeping personal exposure times as short as possible and the amount of radioisotope as low as possible. When working with isotopes such as ^{32}P and ^{125}I it is essential to use shielding; Perspex for the former and lead for the latter. Given the risks associated with using radioisotopes, it is not surprising that many laboratories are now using non-isotopic methods for hybridization studies.

4.3.2 Hybridization conditions

The formation of the duplex structure between the probe and the target sequence is very sensitive to a number of factors especially temperature. Too high a temperature will prevent the formation of any hybrid and too low a temperature will allow mismatching between

AVRIL ROBARTS LRC

the probe and target sequences. Hence it is critically important to select a hybridization temperature which allows the probe to form an exact duplex. Hybridization is usually carried out about 5–15°C below the 'melting temperature' (T_m) of the hybrid (the temperature point at which bonding becomes unfavorable). The T_m depends on the length of the hybrid, the G+C content, ionic concentration of cations in the solution and whether a denaturant such as formamide is present in the incubation mixture. When formamide is the denaturant the T_m in °C can be calculated from the equation:

$$T_m = 79.8 + (18.5 \times \log[Na^+]) + (58.4 \times \%G + C) - (820/\text{bp of hybrid}) - (0.5 \times \%\text{formamide})$$

Therefore, when using a probe for the first time, it is usual to try a range of temperatures around the expected optimum depending on the stringency of the hybrid required. Such initial experiments are important when working with heterologous probes, that is when probing for related genes in DNA or when using a probe from one organism to look for an equivalent gene in another organism.

As described in Section 3.5.2, the hybridization temperatures in the presence of formamide for homologous probes are usually about 42°C and, as a guide, about 37°C for heterologous probes.

Membrane hybridization conditions. Membrane hybridization remains one of the most important methods of hybridization analyses. The procedures involve prehybridization with the hybridization solution, which has no labeled probe included, to block the sites on the membrane that will bind labeled probes non-specifically. If this is not done then there are usually problems with high backgrounds which can mask the real hybridization of the probe. The membrane is then hybridized with the labeled probe and the degree of binding is analyzed. Full details of the procedures used for membrane hybridization are given in Section 3.5.2.

Solution hybridization conditions. Solution hybridization has the advantage that the reaction proceeds 5- to 10-fold faster than when one of the components is attached to a membrane. During solution hybridization methods the hybridization of the target sequence and probe occurs in solution. The non-hybridized nucleic acids and hybrids are then separated. The actual choice of analytical method depends on the nature of the experimental protocol; analytical methods include digestion of non-hybridized sequences using nucleases, separation of hybrids on hydroxyapatite or by ultracentrifugation or gel electrophoresis.

Hybridization solutions usually are composed of salts creating an ideal environment for hybridization to occur (e.g. 0.28 M NaCl, 0.05 M

Protocol 4.1: Solution hybridization

1. In this method the hybridization mixture containing both the probe and target sequence is initiated by heating to 70°C for 5 min to denature secondary structures.
2. The reaction mixture is incubated at 42°C or 37°C for 15 min to allow hybrid formation.
3. Hybrids can then be loaded on to a non-denaturing polyacrylamide gel (9% polyacrylamide with 50 mM Tris-borate, 1 mM EDTA running buffer) and electrophoresed for 4 h at 300 V (4°C).
4. Autoradiography of the dried gel is followed by densitometric quantification of band density.

sodium acetate, 4.5 mM $ZnSO_4$ or 150 mM NaCl, 50 mM Tris-HCl pH 7.4 and 1 mM EDTA). Solution hybridization can be used to quantitate specific RNA molecules as well as investigating functional characteristics. For example, quantitation of snRNA can be carried out using end-labeled DNA oligonucleotide probes [2]. The hybrids are then separated by gel electrophoresis.

Riboprobes are commonly used in solution hybridizations which can analyse quantity or function, for example, RNase protection assays which are a common means of determining exon sizes [3]. The procedure for assessing exon sizes using riboprobes is described in Section 4.3.3.

4.3.3 Analysis of the formation of hybrids

Membrane hybridization is usually assessed by autoradiography as radioactive or chemiluminescent signals that can be recorded by exposure of X-ray film to the isotopic or non-isotopically labeled membrane for a suitable period of time. The exposure times vary between methods. Isotopic detection can be achieved in 10 min for abundant targets to 2 days for low-abundancy targets. The nature of the isotope also defines the amount of time necessary for detection, as strong emitters (^{32}P, ^{33}P and ^{125}I) produce signals quicker than weaker emitters (^{35}S). However, weaker emitters produce better resolution of bands because the particles are less penetrating. Chemiluminescent detection usually takes 1 sec to 5 min to produce an image on X-ray film.

The products of solution hybridization must be separated in a manner that can be used to characterize the labeled hybrid. This can be achieved using polyacrylamide gel electrophoresis (Section 3.3.3) separating labeled species with respect to size, followed by autoradiography. This procedure is often used in structural studies where the size of the hybridizing band may be used to predict intron/exon boundaries or RNA termini.

4.3.4 Types of hybridization assays

Six procedures are commonly used to analyze RNA sequences. All of them can also be used quantitatively to determine the relative concentration of an RNA sequence in RNA populations.

Northern blotting. Northern transfer (blotting) can be used to give an estimate of the length of an RNA transcript when compared to markers of known size. It is the equivalent for RNA analysis that the Southern transfer procedure is for DNA analysis. RNA is separated by electrophoresis on a gel under denaturing conditions (Sections 3.3.3 and 3.3.4), transferred to a nitrocellulose or nylon membrane, and fixed to the membrane. Specific RNA species are then detected by hybridization with a radiolabeled probe. This technique is extremely sensitive. The intensity (assessed by densitometric scanning, Section 3.5.2) of the autoradiographic signal is a measure of the concentration of the specific RNA and the migration position of the band is a measure of its size. This method can be used for analyzing total cellular or cytoplasmic RNA.

The gel electrophoresis step is best performed using conditions which disrupt RNA secondary structure. This greatly improves resolution and allows an accurate estimation of the length of the RNA molecule. Marker RNAs of defined size are commercially available from a number of sources (Section 3.4.2) or they can be prepared when needed. Many different denaturing conditions are commonly used (Section 3.3.4), for example, the use of DMSO/glyoxal, formaldehyde and methyl mercuric hydroxide. The formaldehyde and DMSO/glyoxal based systems are usually used for Northern analysis. All methods use toxic chemicals so care should be taken whilst handling gel solutions and running buffers. Denaturing agarose gel electrophoresis is detailed in Section 3.3.4 and a full description of Northern blotting procedures is given in Section 3.5.2.

RNA dot/slot blots. RNA dot blots can be used to give an estimate of the relative abundance of a specific RNA in a RNA population. In this procedure, a known amount of RNA is immobilized on an inert support such as nitrocellulose or nylon, by baking or UV fixation (Section 3.5.2) and the amount of specific RNA is determined by hybridization with a suitable probe. This technique is rapid and very sensitive and can be made semiquantitative if sufficient radioactivity is hybridized to allow

Protocol 4.2: Analysis of RNA using slot blots

RNA must be denatured before loading into the slot blot apparatus (*Figure 4.2*). Total or mRNA may be loaded into the slots at concentrations that can be detected by hybridization.

1. The volume of the RNA sample is made up to 100 µl and 300 µl of load buffer is added (17.6% (w/v)(5.86 M) formaldehyde, 10 × SSC (1.5 M NaCl, 0.15 M sodium citrate adjusted to pH 7.0 with 10 M NaOH).
2. The RNA is heat-denatured at 65°C for 15 min.
3. Samples are then loaded into the slots.
4. Samples are gently sucked through the nitrocellulose or nylon membranes under vacuum, leaving the RNA on the membrane.
5. Once the wells are empty 200 µl aliquots of wash solution (10 × SSC) are added which are sucked through the membrane, removing contaminants.
6. RNA can then be fixed to the membrane by baking it at 80°C for 2 h or by use of UV cross-linking (Section 3.5.2).

Hybridization of RNA slot blots is usually carried out using formamide-based solutions (*Table 4.2*) at temperatures suitable for the probe:target hybrid (Section 4.3.2) (~42°C). Prehybridization for at least 2 h should precede addition of the probe.

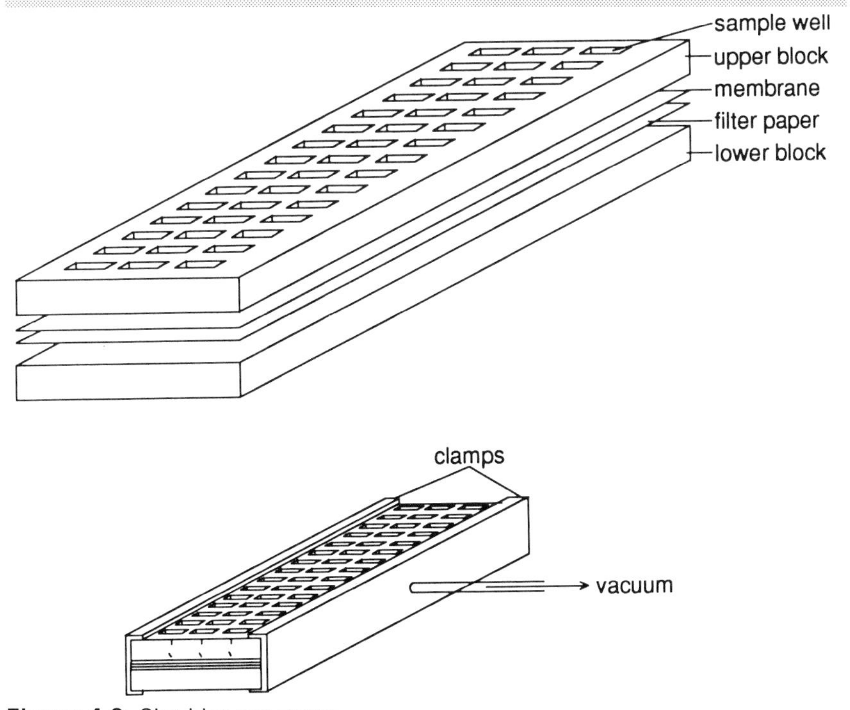

Figure 4.2: Slot blot apparatus.

Table 4.2: Selected hybridization buffers

Constituents
1. 50% formamide, 5 × SSPE[a], 2 × Denhardt's[b], 0.1% SDS, 0.1 mg/ml herring sperm DNA[c]
2. 50% formamide, 5 × SSPE, 5 × Denhardt's, 0.5% SDS, 0.1 mg/ml herring sperm DNA

[a]20 × SSPE stock solution: 3 M NaCl, 0.177 M NaH_2PO_4, 0.02 M EDTA adjusted to pH 7.4 with 10 M NaOH.
[b]100 × Denhardt's: 500 ml: 10 g Ficoll, 10 g polyvinylpyrrolidone, 10 g BSA.
[c]Herring sperm DNA is given here as the blocking agent. This can be replaced with many other DNA sources such as salmon sperm DNA and calf thymus DNA.

scintillation counting of the radioactive 'dots'. Alternatively, densitometric scanning of the X-ray film after autoradiography can also provide a semiquantitative measure of the amount of a particular RNA sequence present. However, dot blots do not provide information as to the number or size of RNA species hybridizing to the probe; for example, if the probe sequence is shared by more than one type of RNA. In addition, it is not possible to obtain information on the rate of transcription only from the amount of transcript that is present.

The probes used for RNA slot blots are usually generated using the random primed hexanucleotide method (Section 3.5.1) and heat-denatured (100°C for 5 min and snap-cooled) prior to addition. Unbound probe is removed by a series of wash procedures detailed in Section 3.5.2.

Nuclease S1 mapping. Nuclease S1 mapping can be used to determine the positions of the 5′- and 3′- termini of a gene and of any introns within it. The enzyme nuclease S1 degrades single-stranded DNA and RNA but at low temperature and in high ionic strength buffers it does not digest double-stranded nucleic acids to any appreciable extent (*Figure 4.3*). During S1 mapping, RNA is hybridized to a single-stranded DNA probe which is complementary to the RNA over only part of its sequence including the 5′ or 3′ termini. After solution hybridization, the reaction mix is incubated with nuclease S1 which degrades unhybridized segments of the DNA probe to leave discrete DNA fragments. The fragments may then be separated with respect to size by polyacrylamide gel electrophoresis. The size of these fragments is equal to the length of the nucleotide sequence over which there is perfect homology between the RNA and DNA. This method was initially used to determine the location and size of intervening sequences in eukaryotic mRNA. The DNA fragments can also be resolved by agarose gel electrophoresis and individual fragments identified by Southern transfer blotting.

A more sensitive and accurate adaptation of the original procedure is to use an end-labeled DNA restriction fragment as a probe. This technique is more technically demanding but it is now more commonly

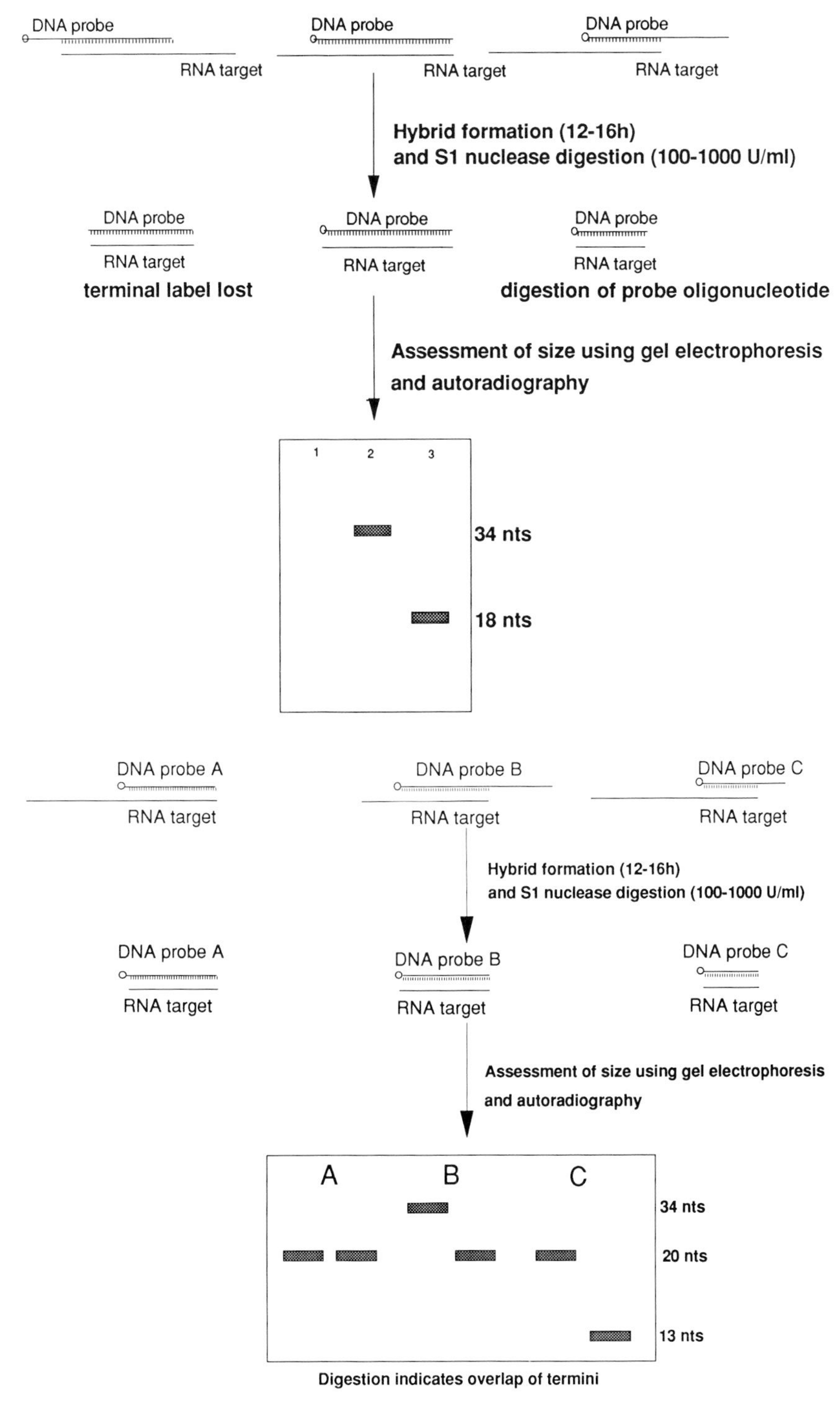

Figure 4.3: Nuclease S1 analysis of RNA using (top) labeled restriction fragments and (bottom) labeled oligonucleotides. Probe sizes before and after digestion are assessed by comparison to ladder fragments (not shown).

used. A restriction fragment of DNA that spans the predicted RNA terminus or splice point is purified and end labeled with ^{32}P as described in Section 3.5.1. This DNA probe is then denatured and hybridized with the RNA and the resulting RNA–DNA hybrids are treated with nuclease S1. The size of the labeled DNA strand in the protected hybrid is then determined by polyacrylamide or agarose gel electrophoresis under denaturing conditions. This corresponds to the distance between the labeled end of the DNA restriction fragment and the RNA terminus or splice point (*Figure 4.4*). Moreover, under conditions of DNA probe excess, the intensity of the autoradiographic signal from this labeled DNA fragment is directly proportional to the abundance of the hybridizing RNA species.

RNA protection assays/RNA mapping. RNA protection assays can provide information as to exon sizes. This type of analysis can be

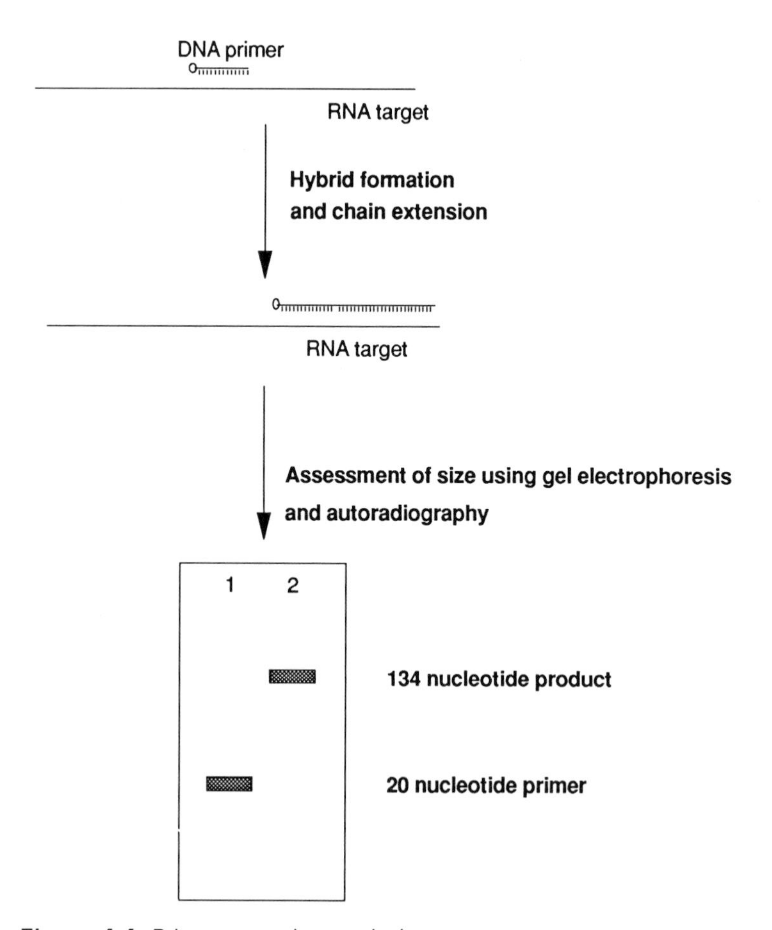

Figure 4.4: Primer extension analysis.

Protocol 4.3: Nuclease S1 analysis of RNA

1. Single-stranded DNA probes are 5′-end labeled with polynucleotide kinase (Section 3.5.1). The concentrations of labeled DNA and RNA used in the assay depend on the abundance of the mRNA species in the sample. Low-abundance mRNA sequences should be assayed by use of high concentrations of sample RNA ($\simeq$250 µg). High-copy-number RNA sequences can be investigated using as little as 0.5 µg sample RNA. The ratio of DNA to RNA is then determined by the size of the DNA probe. Large DNA fragments (e.g. 5 kb) should be present in high concentration (to produce any eventual signal whilst smaller fragments (higher specific activities) can be used in proportionately smaller amounts. Typically, 0.5–250 µg target RNA should be hybridized to 0.1–1.0 µg end-labeled DNA probe.
2. The DNA and RNA are precipitated by adding 0.1 vol. 3 M sodium acetate (pH 5.2) and 2.5 vol. ethanol at −20°C for 30 min.
3. The nucleic acids are pelleted by centrifugation at 15 000 *g* for 15 min at 4°C and washed with ice-cold 70% (v/v) ethanol.
4. The pellet is dried at room temperature and redissolved in 30 µl hybridization buffer (80% (v/v) deionized formamide, 40 mM PIPES-KOH pH 6.4, 1 mM EDTA, 0.4 M NaCl).
5. Heat denaturation is carried out at 85°C for 10 min before immediate transfer to a water bath at the hybridization temperature.
6. The hybridization temperature varies with hybrid G:C content [4]. *Table 4.3* details some hybridization temperatures suitable for hybrids of different G:C contents.
7. Hybridization is carried out over 12–16 h after which 300 µl ice-cold nuclease S1 mapping buffer (0.28 M NaCl, 0.05 M sodium acetate, 4.5 mM $ZnSO_4$ and 100–1000 U/ml nuclease S1) is added.
8. Nuclease S1 digestion of non-complementary nucleic acid occurs over 1–2 h at 0°C before reaction termination.
9. The reaction is stopped by placing on ice and adding 80 µl stop mixture (4 M ammonium acetate, 50 mM EDTA, 50 µg/ml carrier RNA (e.g. yeast tRNA).
10. Deproteinization is achieved with a single phenol–chloroform–isoamyl alcohol extraction followed by addition of 2 vol. ethanol and transfer to −20°C to precipitate the hybrid.
11. Centrifugation at 12 000 *g* for 15 min at 4°C pellets the hybrid which is dried under vacuum at room temperature. The pellet may then be redissolved in 40 µl TE (pH 7.4).
12. The size of the hybrids may then be assessed using agarose gel electrophoresis followed by Northern blotting and autoradiography.

13. Radioactive end-labeled DNA ladder should be co-run with the samples and used to assess the size of the hybrid.

Table 4.3: Hybridization temperatures as defined by the percentage G + C content of the hybrid

% G+C content	Hybridization temperature (°C)
40	48
45	52
50	54
55	59
60	61

carried out before sequencing (preliminary investigation) or after sequencing (supporting evidence for hypothesized exon sizes). The basis of the assay is to protect an exon by hybridization to a probe before digestion of non-protected RNA with various RNases which only digest single-stranded RNA molecules.

Protocol 4.4: RNA protection assays

Riboprobes are prepared by *in vitro* transcription as detailed in Section 3.5.1. Riboprobes for this method are generated from 1 μg linearized template, incorporating [α-^{32}P]UTP (29.6 TBq (800 Ci)/mmol). DNA templates are digested with DNase I and the radiolabeled transcript purified by passage through a Sephadex G-50 (Pharmacia-LKB) spin-column. The optimum conditions for RNA protection assays have recently been determined [3].

1. A concentration of 5–40 μg total RNA in 29 μl hybridization buffer (80% (v/v) recrystallized and deionized formamide, 40 mM PIPES-KOH (pH 6.7), 400 mM NaCl, 1 mM EDTA) is hybridized to 1 μl (5×10^5 c.p.m.) of radiolabeled [^{32}P]RNA transcript.
2. The mixture is heated to 85°C for 5 min and hybridized at 45 or 50°C for 18 h.
3. RNase digestion of non-protected RNA is achieved by addition of 300 μl RNase mixture (10–80 μg/ml RNase A, 0.5–4 μg RNase T1 in 10 mM Tris-HCl (pH 7.5), 5 mM EDTA, 300 mM NaCl) and incubating at 30°C for 10–15 min. Excessive digestion with RNases can produce bands that are smaller than the actual protected sequence.
4. The solution is then deproteinized by digestion with 50 μg/ml Proteinase K (Boehringer Mannheim) and 0.5% (w/v) SDS at 37°C for 10 min followed by a phenol–chloroform–isoamyl alcohol extraction and coprecipitation with 10 μg yeast RNA (heterologous RNA) by addition of 1/10 vol. 3 M sodium acetate (pH 5.6) and 2.5 vol. ethanol.

5. The RNA is pelleted and resuspended in 8 μl loading buffer (7 M urea, 89 mM Tris-borate (pH 8.3), 10 mM EDTA, 0.1% (w/v) bromophenol blue, 0.1% (w/v) xylene cyanol).
6. The hybrids must be heat-denatured before electrophoresis. This is achieved by heating the sample at 90°C for 5 min and snap-cooling in ice. A 6% acrylamide/6 M urea polyacrylamide sequencing gel (12 ml acrylamide stock (30 g acrylamide, 1.5 g bisacrylamide made up to 200 ml with distilled water), 6 ml 5×TBE electrophoresis buffer (0.445 M Tris, 0.445 M boric acid, 12.5 mM EDTA), 10.8 g urea made up to 30 ml with double-distilled water) is used to separate the labeled fragments. Sequencing gel plates are about 50 cm in length and should be cleaned with detergent, ethanol and water before each use. The 'notch' plate should be siliconized by sprinkling dimethyldichlorosilane over the plate and spreading with a tissue in a fume cupboard. Care should be taken as dimethyldichlorosilane is highly toxic. The plates are then 'siliconized' by rinsing with water. Gel mixes can be polymerized by addition of 250 μl 10% (w/v) ammonium persulfate (APS) and 25 μl TEMED. Gel mixes are pipetted down the side of the plates which are separated by 1 mm thick spacers. Bubble formation can be prevented by tapping the border of the rising gel mix as it is poured.
7. Electrophoresis at 1800 V, 40 W for 2 h separates the labeled fragments. The gel plates are separated, leaving the gel attached to the 'back' plate. The gel can be fixed by immersing the plate and gel in 10% (v/v) acetic acid, 10% (v/v) methanol for 15 min. The gel is then transferred to filter paper by laying a sheet over it and peeling the paper back, removing the gel from the plate on to the paper. The gel is dried and exposed to X-ray film at −70°C.
8. The analysis of the exon sizes is accomplished by comparison with marker DNA (e.g. radiolabeled *Hpa*II fragments of a digest of plasmid pGEM2).

Primer extension analysis. Primer extension can be used to determine the position of the 5′ terminus of a gene and to identify the transcripts of related genes. It is the converse of nuclease S1 mapping. A radiolabeled primer derived entirely from within the gene is hybridized to complementary RNA and extended using the enzyme reverse transcriptase. The probe sequence is normally derived from a region near the 5′ end of the gene and the extension reaction terminates at the extreme 5′ end of the RNA (Figure 4.4). As with S1 mapping, this technique can be used to determine precisely the start point of transcription of an mRNA sequence.

The primer extension reaction is more sensitive and yields cleaner results using poly(A)$^+$ RNA. With total RNA there are a relatively large number of prematurely terminated transcripts (i.e. there are many primer extension products which are longer than the probe but which do not extend to the true 5′ terminus of the mRNA). However, with a probe derived from near the 5′ end of the mRNA (i.e. within about 200 nucleotides), perfectly acceptable results can be obtained using total RNA. Thus, poly(A)$^+$ RNA need only be prepared if the level of prematurely terminated transcripts is unacceptable.

Protocol 4.5: Primer extension

1. The DNA primer (usually an oligonucleotide) is 5′-end labeled (Section 3.5.1) and 10^4–10^5 c.p.m. labeled primer is mixed with 0.5–150 µg RNA.
2. The nucleic acid is precipitated with 0.1 vol. 3 M sodium acetate (pH 5.2) and 2.5 vol. ethanol at −20°C for 30 min.
3. Centrifugation at 12 000 *g* for 10 min at 0°C pellets precipitated nucleic acid which is washed with 70% ethanol.
4. The nucleic acid is left to air-dry at room temperature before being redissolved in 30 µl hybridization buffer (80% (v/v) deionized formamide, 40 mM PIPES-KOH, pH 6.4, 1 mM EDTA, 0.4 M NaCl).
5. The nucleic acids are heat-denatured at 85°C for 5 min to remove secondary structure before being placed in a water bath at the annealing temperature. Note that heating samples in microcentrifuge tubes leads to an increase in pressure which can cause some tubes to pop open, releasing aerosols from the radioactive sample. This can be avoided by use of screw-threaded microcentrifuge tubes.
6. The hybridization temperature should be high enough to prevent mismatches forming whilst still allowing accurate duplex formation. Oligonucleotides of between 30 and 40 bases are usually annealed at 30°C whilst double-stranded DNA primers are annealed at 40–50°C.
7. Hybridization is carried out for 8–12 h.
8. The reaction is terminated by precipitation. 100 µl water and 400 µl ethanol are added and the mixture placed on ice for 1 h.
9. Centrifugation at 12 000 *g* for 15 min at 0°C pellets the nucleic acids which are washed with 70% (v/v) ethanol.
10. Residual ethanol is allowed to evaporate at room temperature before the pellet is redissolved in 20 µl reverse transcriptase buffer (50 mM Tris-HCl (pH 7.6), 60 mM KCl, 10 mM $MgCl_2$, 1 mM each dNTP, 1 mM DTT, 1 U/µl placental RNase inhibitor, 50 µg/ml actinomycin D).
11. Reverse transcription is initiated by addition of 50 U of

Moloney leukemia virus (M-MLV, Gibco–BRL) reverse transcriptase and incubation at 37°C for 2 h.

12. The reaction is terminated by addition of 1 μl 0.5 M EDTA (pH 8.0) and 1 μl DNase-free pancreatic RNase (5 μg/ml).
13. The RNA portion of the hybrid is then removed by digestion at 37°C for 30 min.
14. Digestion is terminated by addition of 150 μl 0.1 M NaCl in TE and extracted with an equal volume of phenol–chloroform–isoamyl alcohol.
15. 500 μl ethanol is added to the recovered aqueous layer and placed on ice for 1 h (ethanol precipitation).
16. Centrifugation at 12 000 *g* for 15 min at 0°C pellets the nucleic acid which is washed with ice-cold 70%(v/v) ethanol. Residual ethanol is allowed to evaporate at room temperature before the pellet is redissolved in 4 μl TE (pH 7.4).
17. The sizes of the extended strands may then be determined by polyacrylamide gel electrophoresis: 6 μl loading buffer (80% formamide, 10 mM EDTA, 1 mg/ml bromophenol blue, 1 mg/ml xylene cyanol) is added to the sample. The nucleic acid is heat-denatured at 85°C for 5 min and snap-cooled in ice to prevent reannealing. Electrophoresis through a 7 M urea/polyacrylamide sequencing gel with radiolabeled DNA markers allows size assessment of the extended primer.

***In situ* hybridization.** *In situ* hybridization, is hybridization of a nucleic acid probe to nucleic acids within cytological preparations, permitting the spatial localization of sequences complementary to the probe. This localization can be useful in a number of ways for answering biological questions. It allows a precise analysis of the tissue distribution of any RNA species of interest. In addition, *in situ* hybridization makes it possible to study the RNA of individual cells unaffected by the RNA of other cells in the tissue. Thus it is possible to use the technique to detect RNAs that are present in only very small populations of cells. Such RNAs might never be detected in RNA extracted from whole tissue because of dilution by other RNA species from the vast majority of the cells which do not contain the RNA of interest. Thus, *in situ* hybridization may be useful not only to show where an RNA is localized but, in some cases, it may be the best way to show that the RNA exists at all. *Figure 4.5* shows an example of *in situ* hybridization.

Most studies of the localization of RNA in cytological preparations require preservation of the morphology of the entire cell and, in some cases, of tissues or organs. Many conventional techniques for fixation are compatible with hybridization to cellular RNA so it is worthwhile

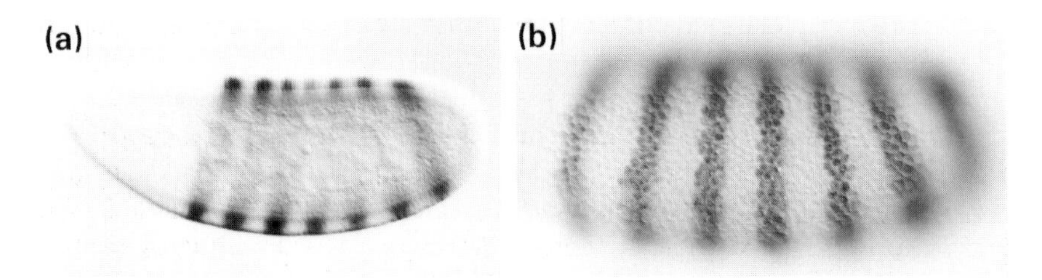

Figure 4.5: RNA:RNA *in situ* hybridization to whole-mount *Drosophilia* embryos at (a) low and (b) high resolution showing expression patterns of the segmentation gene *fushi tarazu (ftz)*. The probe was labeled with digoxigenin and detected using the alkaline phosphatae-catalyzed precipitation of 5-bromo-3-chloro-3-indolylphosphate/nitroblue tetrazolium (BCIP/NBT). Expression occurs in seven transverse stripes separated by stripes of non-expressing cells in a 2.5-h-old wild-type embryo. The probe demonstrates that the pattern of segmentation becomes established in early embryo development. Reproduced from Ref. 5.

experimenting with several different techniques to find the one most suitable to the biological material of interest. The best cytological techniques for hybridization to cellular RNA should: (i) preserve the morphology of the cells; (ii) not extract or modify the RNA; (iii) not change the location of the RNA; (iv) leave the RNA accessible to the hybridization probe.

Several preparation procedures can be followed [5], which are based around the following procedural points.

1. The tissue of interest is frozen and embedded.
2. The embedded tissue is then cut into 5 μm thick sections and placed on subbed slides.
3. The samples are dried for 2 days at 40°C before acetylation with acetic anhydride.
4. Since the target is single-stranded RNA, the most effective method is to use a single-stranded probe complementary to the target RNA. This is most easily prepared by constructing an appropriate SP6 recombinant which will then produce antisense RNA (Section 3.5.1).
5. Hybridization to cellular RNA is usually carried out in the presence of formamide to prevent degradation of cellular RNA. It should be noted that formamide reduces the melting point of RNA–RNA duplexes by 0.35°C for every 1% formamide.

4.4 RNA fingerprinting

Although RNA fingerprinting has declined in popularity as RNA sequencing technology has become simpler, faster and more automated, it remains an extremely useful technique for detecting sequence similarities between different species of RNA and is often the fastest way to obtain accurate information about the reactions associated with RNA processing.

Several different methods can be used to prepare RNA fingerprints, which are two-dimensional arrays (Section 3.3.5) of oligonucleotides derived from partial digestions with ribonucleases. All methods begin with ribonuclease digestion. *Table 4.4* lists the specificities of the nucleases used most frequently for RNA fingerprinting; note that the specificity is much more limited than those of restriction nucleases used to produce specific fragments of DNA. Usually either RNase T1, which cleaves after G residues, or pancreatic RNase A, which cleaves after C and U residues, is used for this step.

1. Ribonuclease digestion is carried out in solutions appropriate for the nucleases. One frequently used solution is 10 mM Tris-HCl (pH 7.5), 5 mM EDTA, 300 mM NaCl incubating with 0.1–4 μg of the appropriate RNase at 30°C for 1 h.
2. The solution is then deproteinized by digestion with 50 μg/ml Proteinase K (Boehringer Mannheim) and 0.5% (w/v) SDS at 37°C for 10 min followed by a phenol–chloroform–isoamyl alcohol extraction. The recovered supernatant is ethanol precipitated by addition of 1/10 vol. 3 M sodium acetate and 2.5 vol. ethanol.

After RNase digestion, the two-dimensional arrays (fingerprints) can be separated using cellulose acetate strips followed by DEAE paper electrophoresis, or by gel electrophoresis. Gel electrophoresis has become the favored method for fingerprinting but both procedures are detailed in the following sections.

Table 4.4: Specificities of nucleases for RNA sequences

Ribonuclease	Specificity
RNase T1	GpN
RNase U2	ApN
RNase CL3	CpN
RNase Phy M	UpN and ApN
RNase *B.cereus*	UpN and CpN

A, adenosine; C, cytosine; G, guanosine; U, uracil; N, any nucleoside.

4.4.1 Fingerprinting using cellulose acetate–DEAE paper electrophoresis [6]

In this method the oligonucleotides are analyzed on the basis of charge in the first dimension of electrophoresis and on the basis of size in the second dimension.

Protocol 4.6: Fingerprinting using cellulose acetate–DEAE paper electrophoresis

First dimension

1. Cellulose acetate (CA) strips are used for the first dimension. Fingerprinting on CA strips is fast (20 min for electrophoresis and 20 min for transfer to the diethylaminoethyl (DEAE) thin-layer plate used for the second dimension). Under ideal conditions, eight RNA digests can be processed in one batch.
2. The CA strip is soaked in electrophoresis buffer containing 0.3 M ammonium formate/7 M urea (pH 3.5 with pyridine). The CA strip is balanced on two glass rods and the middle of the strip is gently blotted dry before sample addition.
3. The sample is spotted on to the middle of the strip and the dye mixture (2% (w/v) fuchsin, 1% orange C, 1% xylene cyanol) is applied to the strip on either side of the sample.
4. The strip is placed in the electrophoresis tank and electrophoresed at 50–100 V cm^{-1} for 20 min.

Second dimension

Oligonucleotides are usually transferred to DEAE-cellulose thin-layer plates and ascending RNA homochromatography used for the second dimension. Homochromatography is the separation of radiolabeled molecules by displacement from the DEAE groups by saturated addition of a homomixture containing non-labeled RNA molecules.

1. The CA strip holding the first dimensional separation is laid on to a DEAE plate (45–100 cm). The oligonucleotides are transferred to the DEAE by placing a pad of five water-soaked (1 × 50 cm) Whatman No. 3 MM paper strips on to the CA strip and pressing evenly for 2 min.
2. Prechromatograph with water allowing migration for 15 cm at 70°C before replacing water with electrophoresis buffer (2.5% (w/v) formic acid, 8.7% (v/v) acetic acid, pH 1.9 or 6.5% (w/v) formic acid, 0.1 M pyridinium formate, pH 2.3).
3. Electrophoresis is carried out at 30 V/cm for short DEAE papers and 20 V/cm for long DEAE papers for 16–24 h (depending on buffer, oligonucleotide composition and paper size). Migration is stopped when the blue dye has nearly reached the edge of the plate.
4. The plate is dried and marked (for orientation purposes) with radioactive ink before autoradiography.
5. If it is necessary to analyze the oligonucleotides further, they can be eluted from the DEAE plate with relative ease. The

radioactive areas can be located by placing the developed autoradiograph under the DEAE-cellulose plate and illuminating from below. The outlined areas can be sucked into a drawn-out glass tube (0.3 cm × 4 cm under vacuum). The tube can then be placed into a punched microcentrifuge tube which will act as a container for RNA elution. RNA is eluted with three sequential elutions by centrifugation at 750 *g* for 2 min with 80 µl 1 M NaCl in a swinging bucket rotor. RNA can then be ethanol precipitated.

Polyethyleneimine (PEI) plates can also be used to obtain second-dimensional separations which are similar to those obtained with DEAE-cellulose.

4.4.2 Fingerprinting using a two-dimensional separation of RNA fragments on polyacrylamide gels [7]

Separation in the first dimension can be run on a polyacrylamide gel without specialized apparatus and this method gives more reproducible patterns than the cellulose acetate method. However, the first-dimensional gels are time-consuming (2 h for electrophoresis and 2 h for transfer), and only four RNA digests can be fractionated in each 20 × 40 cm gel. First-dimensional gels usually yield excellent RNase T1 fingerprints of ^{32}P-labeled oligonucleotides (both internally and end labeled), but can sometimes be less reliable with RNAs digested by pancreatic RNase A.

Protocol 4.7: Fingerprinting using gel electrophoresis

First dimension

1. In the first dimension a 10% acrylamide/6 M urea gel (pH 3.3) is used to separate the oligonucleotides on the basis of base composition as well as chain length. The denaturant, urea, ensures that oligonucleotides do not aggregate. The gel mixture is citric acid-based and can be prepared by mixing 37.5 ml 40% (w/v) acrylamide/1.3% (w/v) bisacrylamide, 3.75 ml 1 M citric acid, 100 ml 9 M urea, 0.6 ml 2.5% (w/v) $FeSO_4.7H_2O$, 0.6 ml 10% (w/v) ascorbic acid and 60 µl 30% hydrogen peroxide.
2. Radiolabeled RNA digest is mixed with 100 µg carrier RNA/ 1–20 × 10^6 c.p.m. of digest and is loaded in gel-loading buffer (electrophoresis buffer supplemented with 10–20% (w/v) sucrose, 0.25% (w/v) bromophenol blue, 0.25% (w/v) xylene cyanol FF).
3. Electrophoresis is carried out using a 0.025 M citric acid, 6 M urea electrophoresis buffer at 500 V for 20 h or 900 V for 6 h (until bromophenol blue dye has moved about 20 cm).

Second dimension

1. The second-dimension gel has a higher acrylamide concentration of 20–22%, contains no urea and is buffered to pH 8.0 or 8.3. The gel is made up with electrophoresis buffer (Tris-citrate (detailed above) or Tris-borate (Section 3.3.1)).
2. The first-dimension separation is mounted in the second-dimension gel as detailed in Section 3.3.5.
3. Electrophoresis at 350 V for 15 h separates the radiolabeled fragments.
4. Autoradiography can then be used to visualize the 'fingerprint'.

High resolution can be achieved by the use of tube gels in the first dimension. This, however, limits the number of samples per second-dimensional gel to only one.

Polyacrylamide gels can produce very good second-dimensional separations and are used by a number of groups; homochromatography, however, has a greater ability to separate oligonucleotides evenly over a wide range of sizes.

This fingerprint technique can easily be applied, not only to tRNA genes and their *in vivo* or *in vitro* transcripts, but also to more complex genes such as viral genomes (*Figure 4.6*). The interpretation of fingerprints is based on the fact that each oligonucleotide is characterized by its location on the fingerprint, depending on its length and base composition as follows. U contributes to a high mobility in the first dimension and this contribution decreases in the order G, A and C. In the second dimension, shorter oligonucleotides move faster than longer ones although all four bases contribute individual mobilities. If the identity of oligonucleotides is questionable, they can be eluted from the DEAE-cellulose plate in order to characterize the nucleotide neighboring the 3′ terminal guanosyl nucleotide by RNase T2 digestion and thin-layer chromatography (TLC) or, more definitively, the fragment can be sequenced.

4.5 Direct sequencing methods for RNA

Chemical sequencing methods were devised originally for the sequencing of RNA by sequential cleavage of the nucleotide chain using periodate which cleaves at the 3′ position of the ribose ring, followed by selective removal of the oxidation products. However, this procedure only works well with short oligonucleotides and is affected by ribose methylation which is quite common in some types of RNA, for example, rRNA and tRNA. This chemical degradation method is no

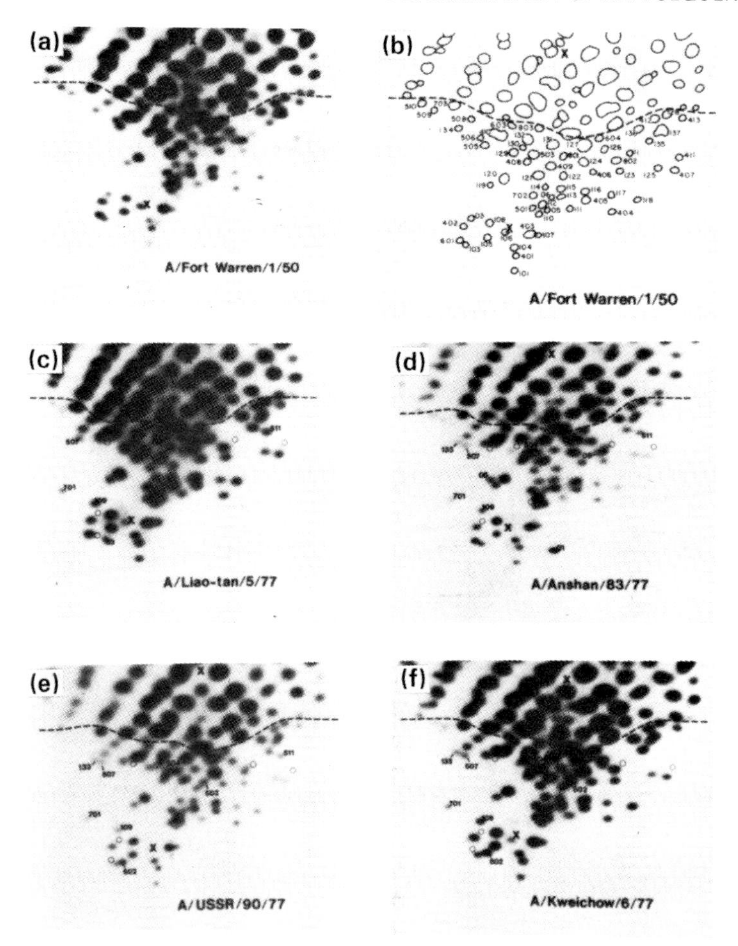

Figure 4.6: Fingerprints of RNase T1 digests of influenza virus RNAs. The viral RNAs isolated from different H1N1 influenza A virus variants were digested with RNase T1 and the resulting oligonucleotide mixture was 5′-end labeled using [γ-^{32}P]ATP and polynucleotide kinase. Only the spots below the dashed line are used in comparing different viral variants, and these are characterized by gel sequencing methods. As compared to the reference strain (a, b), the variants (c–f) show additional spots, indicated by new numbers, and missing spots, indicated by circles. The origin is at the bottom left of each autoradiograph. Xs indicate the position of the dye spots xylene cyanol FF (bottom left) and bromophenol blue (top middle). Reproduced from Young *et al.* (1979) *Cell,* **18,** 73, with permission from Cell Press.

longer used now that faster methodologies analogous to those used for DNA have become available.

Current RNA sequencing procedures employ two different methods. These are a nuclease digestion method which is based on a similar principle as the Maxam–Gilbert method of partial specific degradation for DNA sequencing or using reverse transcriptase to produce a cDNA which is sequenced using the Sanger dideoxy chain termination method. Commercial kits are available for both of these sequencing techniques.

4.5.1 Sequencing RNA using partial nuclease degradation

This method uses base-specific ribonucleases (*Table 4.5*) to degrade end-labeled RNA partially and the labeled fragments are then separated on a sequencing gel and analyzed by autoradiography. The

Table 4.5: Solutions used for partial RNA digestion

Ribonuclease	Digestion solution	Specificity
RNase T1	8 M urea, 2 mM EDTA, 20 mM sodium citrate (pH 3.5)	GpN
RNase U2	8 M urea, 2 mM EDTA, 20 mM sodium citrate (pH 3.5)	ApN
RNase CL3	8 M urea, 20 mM Tris-HCl (pH 7.5)	CpN
RNase Phy M	8 M urea, 1 mM EDTA, 20 mM sodium citrate (pH 5.0)	UpN and ApN
Sulfuric acid	7 M urea, 0.11 M sulfuric acid	NpN
Nuclease S7	8 M urea, 20 mM Tris-HCl (pH 7.5)	NpU and NpA

A, adenosine; C, cytosine; G, guanosine; U, uracil; N, any nucleoside.

Protocol 4.8: Partial nuclease digestion of RNA

1. End-labeled RNA (5′ or 3′) is divided into six tubes and four specific ribonuclease and one non-specific chemical cleavage reaction are set up (T1, U2, S7, sulfuric acid, CL3). The sixth tube contains an alkaline hydrolysis buffer. The solutions used for partial ribonuclease digestion are detailed in Table 4.5.
2. The six tubes are then incubated at 50°C for 30 min. The incubation conditions can be modified depending on the length of the RNA to be sequenced.
3. The tubes are cooled and the contents of each of the tubes is loaded into separate tracks of a sequencing gel (e.g. 15–20% acrylamide containing 8 M urea in Tris-borate–EDTA electrophoresis buffer, Section 3.3.1).
4. After electrophoresis the gel is dried down and autoradiographed to reveal the bands of the fragments of the RNA. An example of the sequence generated is illustrated in Figure 4.7.
5. Sequencing gels are always read from the base of the gel; these smaller fragments can represent a known sequence or may just provide an easy start to analysis as these bands are often easiest to visualize. The products of the ribonuclease T1 and U2 reveal G and A residues whilst the CL3 track represents C residues. Uracil residues are delineated in the S7 lane at a position one before their true place in the sequence. This is due to cleavage occuring before the uracil residue. Thus, reading up the gel the sequence can be recorded. The hydrolysis tract (sulfuric acid) contains all the nucleotides and is an important control. One of the major weaknesses of this method is that the RNA must be pure, in contrast with the dideoxy method where purity is less critical.

principle of the method is shown in *Figure 4.7.* Using this type of approach 40–60 nucleotides can be sequenced at a time.

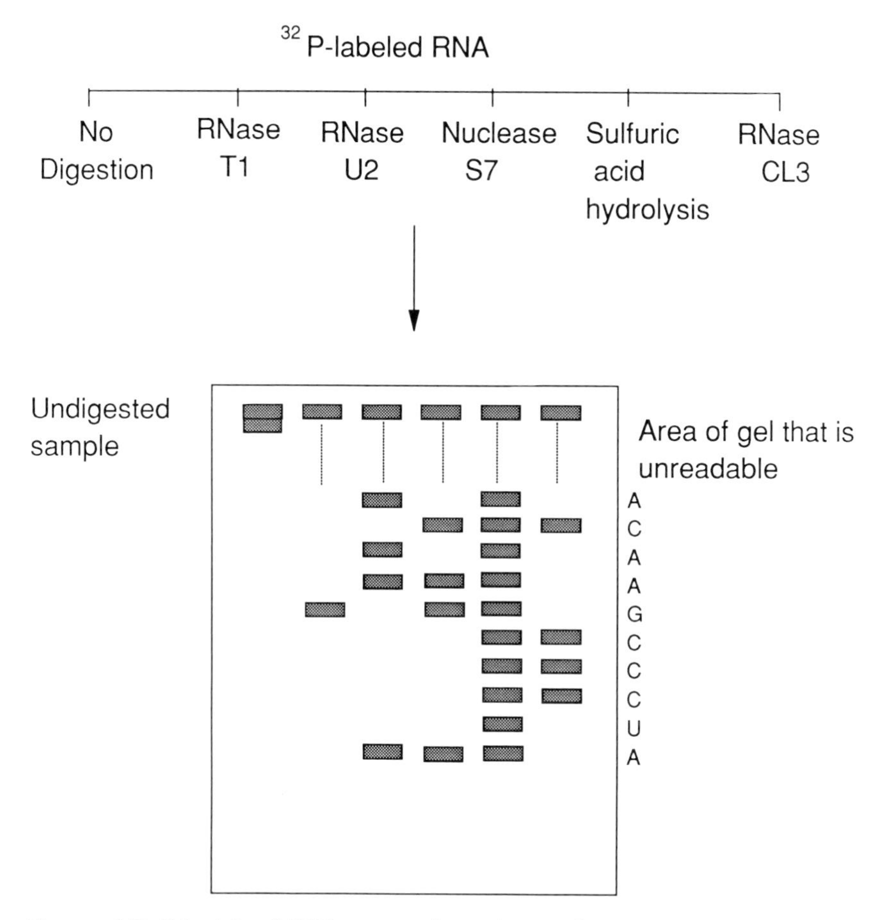

Figure 4.7: Principle of RNA sequencing using nucleases.

4.5.2 Sequencing RNA using the dideoxy chain termination method

Three methods are available to sequence RNA using the Sanger-based termination method during enzymatic copying. The first copies the RNA template, synthesizing a complementary DNA (cDNA) molecule but occasionally terminating at each base position. This produces a banding pattern which may be read as sequence in a similar manner to the above. The second method relies on a complete cDNA molecule being synthesized before sequencing. The complete cDNA can then be sequenced as a DNA molecule avoiding degradation problems inherent in the first method. The last method uses probably the most powerful molecular technique—the PCR (Section 5.5). PCR can be used in conjunction with the enzyme reverse transcriptase to amplify RNA sequences (producing cDNA) followed by exponential amplification of this cDNA. The amplified cDNA can then be sequenced by a PCR-based Sanger-type method. These procedures are now described in more detail.

Sequencing RNA during first-strand cDNA synthesis. The principle of this method is that the RNA can be copied into a labeled strand of cDNA by a viral reverse transcriptase enzyme which extends a short primer molecule that has been hybridized to the specific target RNA sequence (*Figure 4.8*). The primer is a short DNA oligonucleotide that is complementary to a region flanking the sequence under study. Inclusion of chain terminators during cDNA synthesis results in premature termination and generates a ladder of fragments corresponding to the RNA sequence (*Figure 4.8*).

One of the important advantages of this method is that the RNA to be sequenced does not have to be pure; RNA samples in which the template constitutes less than 1% of the total RNA have been sequenced successfully. The reason for this is that the reverse transcriptase can only extend an existing primer molecule which is hybridized to the template RNA. However, obviously it is important that the primer only hybridizes to the RNA to be sequenced and not to any other RNA. Usually the cDNA copy is labeled by the incorporation of a nucleotide labeled with [^{32}P]/[^{33}P]phosphorus or [^{35}S]sulfur; the latter isotope gives better results in terms of resolution and is also less of a radiological hazard.

Protocol 4.9: Sequencing during cDNA synthesis

1. The RNA template (2 µl–0.5 mg/ml) is mixed with DNA primer (2 µl–5–25 µg/ml) and 1 µl 5 × hybridization buffer (0.5 M KCl, 0.25 M Tris-HCl, pH 8.5).
2. The mixture is heat-denatured before hybridization at 90°C for 1 min, allowing the sample to cool slowly at 40°C. This slow cooling allows the primer to anneal to the RNA.
3. The extension reaction is set up by predrying 185 kBq (5 µCi) [α-^{32}P]dATP in a 0.5 ml microcentrifuge tube and adding 5.5 µl hybridized RNA, 5.5 µl 5 × reverse transcriptase buffer (0.25 M Tris-HCl (pH 8.5), 0.25 M KCl, 0.05 M DTT, 0.05 M $MgCl_2$) and 5.5 µl reverse transcriptase (1000 U/ml).
4. This mixture is aliquoted (3 µl) into four tubes which represent the four bases of RNA.
5. Each of the termination mixes (*Table 4.6*) that are then added immediately contain not only the normal four 2′-deoxynucleotides (dATP, dCTP, dGTP and dTTP) needed to synthesize the DNA copy but also a small amount of one 2′,3′-dideoxynucleotide (ddATP/ddCTP/ddGTP or ddTTP) which, when incorpo-

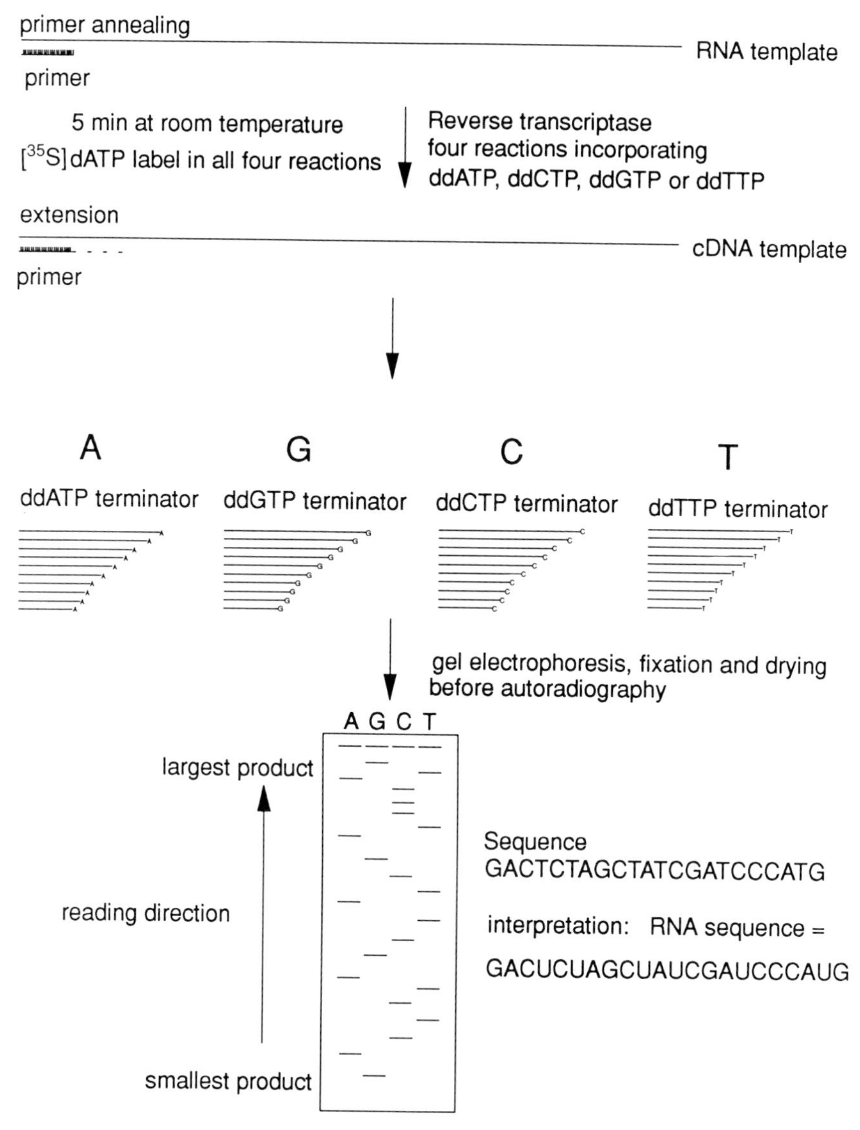

Figure 4.8: Sequencing of RNA during first-strand cDNA synthesis using the dideoxy method.

rated, will block any further chain extension as there is no 3′ hydroxyl group. By setting up four incubation mixtures, each with a different dideoxynucleotide, it is possible to obtain DNA fragments that are randomly terminated at all four nucleotides.

6. Chain extension with occasional termination occurs at room temperature for 5 min followed by incubation at 37°C for 30 min.

7. A chase mixture is added to extend any non-terminated strands to completion. Replication that has not proceeded to completion could produce small fragments of DNA that would appear on the autoradiograph and be interpreted as products of a sequencing reaction. 1 μl of this chase mixture (1 mM each dNTP, 6 mM Tris-HCl (pH 8.5)) is added to each of the four tubes and incubated at 37°C for 15 min.
8. Reaction is terminated by addition of 10 μl stop mix (92% (v/v) formamide, 5 mM EDTA (pH 7.2),0.05% (w/v) bromophenol blue, 0.05% xylene cyanol) to each of the four tubes.
9. The products of the four incubations are then separated on polyacrylamide sequencing gels (6% acrylamide/urea gel mix—*Table 4.7*). Separation by electrophoresis may be enhanced by use of a gradient gel where a strong concentration of buffer (TBE) at the base of the gel slows the migration of smaller molecules. This can also be achieved by increasing the thickness of the gels (wedge gels) at the base, making migration slower at these points. The samples are heat-denatured before loading on to the gel by boiling for 2 min and snap-cooling to prevent reannealing.
10. Electrophoresis at 1600 V, 40 W for 3 h provides a good separation of the radiolabeled fragments.
11. Gels holding ^{35}S- or ^{33}P-labeled cDNA need to be soaked in 10% (v/v) acetic acid, 15% (v/v) methanol for 20 min to remove the urea that is present in the gel before they are dried down for autoradiography. It is not necessary to soak gels of [^{32}P]cDNA as the emission strength is great enough to penetrate wet gels. However, drying gel will increase the resolution of the autoradiography and is to be recommended.
12. Autoradiography is carried out using Kodak XAR-5 film, although better contrast at the expense of a longer exposure time can be achieved using Kodak Ektascan film; other manufacturers produce similar films that can also be used. Gels containing ^{32}P-labeled cDNA should have their exposure times shortened and be used in concert with a tungstate intensifier screen (indirect autoradiography), placing at −80°C.
13. From the positions of the labeled DNA molecules it is possible to read off the sequence of the original RNA; a sequencing gel is shown in *Figure 4.9*. As many as 400 nucleotides can be read from a single sequencing gel. Most sequencing is carried out using radioisotopic labeling, although there is a trend towards the use of non-isotopic labeling procedures which have a similar sensitivity range and avoid the hazards of radioisotopes.

Table 4.6: Termination mixes

Termination mix	Constituents
A	0.28 mM dCTP, dGTP, dTTP, 0.069 mM [α-^{35}S]dATP, 0.012 mM ddATP in Tris-HCl (pH 8.0)
G	0.28 mM dCTP, dGTP, dTTP, 0.069 mM [α-^{35}S]dATP, 0.17 mM ddGTP in Tris-HCl (pH 8.0)
C	0.28 mM dCTP, dGTP, dTTP, 0.069 mM [α-^{35}S]dATP, 0.12 mM ddCTP in Tris-HCl (pH 8.0)
T	0.28 mM dCTP, dGTP, dTTP, 0.069 mM [α-^{35}S]dATP, 0.35 mM ddTTP in Tris-HCl (pH 8.0)

Table 4.7: Sequencing gel mixes

Gel mix	Constituents
Bottom solution: 5 × TBE/6% acrylamide mix	6% acrylamide[a], 5 × TBE[b], 7.67 M urea, 10 mg bromophenol blue
Top solution: 0.5 × TBE/6% acrylamide mix	6% acrylamide, 0.5 × TBE, 7.75 M urea

[a]40% acrylamide mixes consisting of premixed acrylamide and bisacrylamide are commercially available.
[b]10 × TBE stock: 0.89 M Tris base, 0.89 M boric acid, 25 mM EDTA.

The sequencing gel may have one of two well types. The standard gel electrophoresis well formed by removal of an intrusion in the set gel is rarely used as it has been largely superseded by use of 'shark's tooth' combs (*Figure 4.10*). These combs are placed with their flat edge in the gel as it sets forming a flat top to the gel. Before electrophoresis the comb is removed and the teeth are pushed 1 mm into the gel. The teeth isolate a defined area into which the sample may be loaded. This area is delimited by the glass plates and the sides of the teeth on either side. Loading sequencing gels is troublesome as it should be performed quickly. A common problem is found when repeated use of a pipette tip results in blockage, making a change of tip necessary. When using duck-billed tips that fit a Gilson P20 (or equivalent micropipettor) the tip may be modified before loading by cutting the stem at a lower point so that it fits on to a Gilson P10. This will afford greater control during sample expulsion into the well. Fresh tips can be used for each reaction or the tip may be washed by two quick washes in the electrophoresis buffer, keeping the tip in the solution at all times, removing the tip from the buffer with the plunger fully depressed. This ensures that any solution left in the tip is at the extreme base, and can be removed by touching on to an absorbent tissue.

Dideoxy enzymatic chain termination method applied to sequencing of a cDNA template. This method is possible once a complete cDNA molecule has been made. A cDNA molecule can be made from any RNA molecule by annealing a complementary primer to it. cDNA can be synthesized as above including all necessary dNTPs, omitting

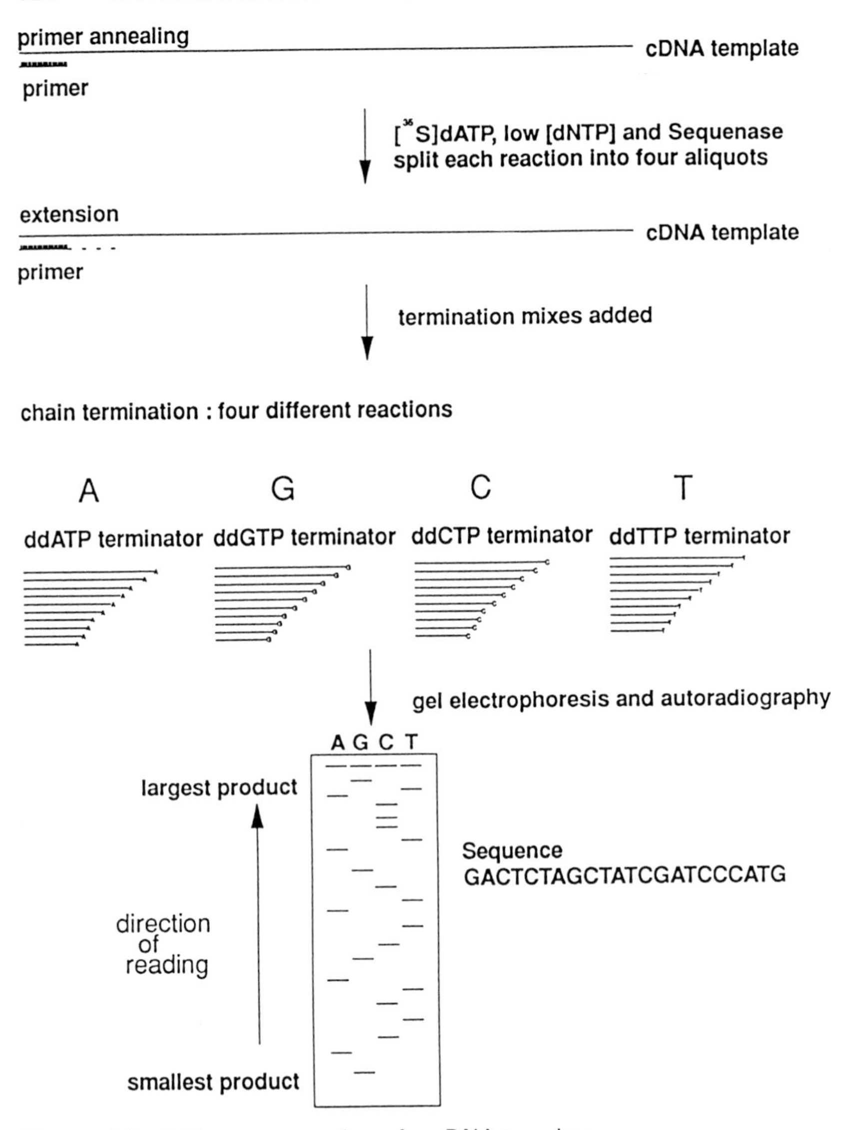

Figure 4.9: Dideoxy sequencing of a cDNA template.

the chain terminating ddNTPs. cDNA can also be synthesized using commercially available kits (Promega) or by using RT-PCR. However, it should be noted that premature termination of the cDNA can occur, particularly when dealing with RNA possessing a high degree of secondary structure, making it important to ensure that the cDNA is of the expected length.

The cDNA may then be sequenced using a primer complementary to the cDNA strand. This primer would be the equivalent of RNA sequence or be complementary to an engineered end of the primer used

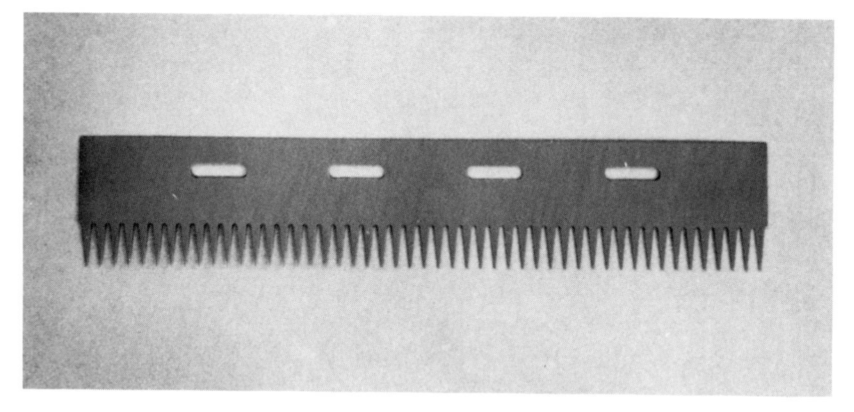

Figure 4.10: Shark's tooth comb for loading samples on to sequencing gels.

in the reverse transcriptase reaction. The theory behind sequencing DNA is similar to the method detailed above in that fragments are generated by occasional chain termination events.

The template (cDNA) concentration for this method of DNA sequencing should be 0.1–0.5 μg/μl. If the cDNA concentration is too low then it can be amplified using the polymerase chain reaction or by cloning into a plasmid vector, transforming *E.coli*, bacterial growth and subsequent recovery of large amounts of plasmid DNA containing the cDNA.

Protocol 4.10: Sequencing of a cDNA template

1. Template denaturation: 20 μl template is taken and alkali-denatured with 5 μl fresh 1 M NaOH, 1 mM EDTA at room temperature for 5 min. The denatured single-stranded DNA is purified by passage through Sepharose Cl6B spin-columns before primer annealing (Section 3.5.1).
2. Primer annealing: 8 μl template is mixed with 1 μl primer (0.3 pmol/μl) and 2 μl reaction buffer (200 mM Tris-HCl (pH 7.5), 100 mM $MgCl_2$, 250 mM NaCl). The primer is allowed to anneal for 15 min at 37°C.
3. Labeling reaction: Four labeling reactions are set up by aliquoting 2.5 μl annealed template solution into tubes or microtiter plate wells and adding 2 μl enzyme mix.
   Enzyme mix:
   66 μl 10 mM Tris/0.1 mM EDTA ($T_{10}E_{0.1}$) pH 7.5
   11 μl 0.1 M DTT
   4.5 μl label mix (7.5 μM Deaza-GTP, 7.5 μM dCTP, 7.5 μM dTTP in $T_{10}E_{0.1}$)
   3 μl (13 U/μl) Sequenase v2.0 (USB)

5 μl [α-^{35}S]dATP (18.5 TBq (500 Ci)/mmol at 370 MBq (10 mCi)/ml).

Deaza-GTP is a nucleotide analog used to allow sequencing through GC-rich areas which sometimes causes band compressions during sequencing. Note siliconized pipette tips should be used as the enzyme adheres to non-siliconized tips.

4. Chain termination: immediately after adding the enzyme mix 2 μl of the termination mixes (A, G, C, T) should be added to the appropriate tubes. Termination mixes are detailed in *Table 4.8*. Chain extension with periodic termination events occurring continues for 5 min at 37°C.
5. Reaction termination: The reaction is terminated by the addition of the sequencing gel loading buffer (98% formamide, 1 mg/ml xylene cyanol, 1 mg/ml bromophenol blue, 0.01 M EDTA.
6. Gel electrophoresis: The samples are heat-denatured before loading on to the gel by boiling for 2 min and snap-cooling to prevent reannealing. The sequencing gel, electrophoretic conditions and post-electrophoretic treatments are identical to those described above.

Table 4.8: Termination mixes

Termination mix	Constituents
A	0.1 mM dATP, dCTP, dGTP, dTTP, 0.01 mM ddATP in $T_{10}E_{0.1}$
G	0.1 mM dATP, dCTP, dGTP, dTTP, 0.01 mM ddGTP in $T_{10}E_{0.1}$
C	0.1 mM dATP, dCTP, dGTP, dTTP, 0.01 mM ddCTP in $T_{10}E_{0.1}$
T	0.1 mM dATP, dCTP, dGTP, dTTP, 0.01 mM ddTTP in $T_{10}E_{0.1}$

PCR sequencing of RNA. This method is probably the quickest way to determine RNA sequences. The system is based on synthesis of a complete cDNA as a preliminary step of PCR [8]. The cDNA is then sequenced using a method based on chain termination by dideoxynucleotides. The advantage of this system is that femtomoles of RNA can be amplified into sequence from micropreps of RNA (Section 2.7) and that manipulation is reduced to a minimum.

The first PCR cycle copies the RNA of interest, from a specific primer, producing a cDNA strand. This strand can then be sequenced.

Protocol 4.11: PCR sequencing of RNA

1. Reverse transcription reaction:
   - 50 pmol primer
   - 1 μl dNTPs (10 mmol each dNTP, Pharmacia)
   - RNA sample
   - 1 μl (0.2 U) reverse transcriptase (BRL)

1 μl (10 U) RNasin (RNase inhibitor, Promega)
1 μl 10× reverse transcription buffer (100 mM Tris-HCl (pH 8.8), 500 mM KCl, 0.1% (w/v) gelatin, 1% (v/v) Triton X-100, 1.5 mM $MgCl_2$)
to 10 μl with sterile water and mixed by vortexing.

2. Transfer to a thermal cycler (Hybaid) and incubate at 42°C for 30 min.
3. Heat-kill the reverse transcriptase at 95°C for 5 min in the thermal cycler.
4. Template purification: Remove unincorporated dNTPs by passing through a Sepharose CI6B spin-column, recovering cDNA in the eluant. The sequencing reaction may then be set up using commercially available PCR sequencing kits (e.g. Sequitherm™ cycle sequencing kit, Cambio).
5. PCR sequencing (Cambio kit): The sequencing primer is 5′ end labeled with (370 MBq (10 μCi) [α-^{32}P]dATP or [α-^{33}P]dATP at 29.6 TBq (800 Ci)/mmol (Section 3.5.1)) before sequencing.
6. Premix:
   1.5 pmol end-labeled primer
   2.5 μl 10× sequencing buffer (0.5 M Tris-HCl (pH 9.3), 25 mM $MgCl_2$)
   10 fmol cDNA for [α-^{32}P]dATP end-labeled primer or 50 fmol for [α-^{32}P]dATP end-labeled primer
   to 16 μl with double-distilled water before adding 1 μl sequiTherm thermostable DNA polymerase.
7. For each template, four 0.5 ml microcentrifuge tubes are labeled (A, G, C, T) and to these 2 μl of the appropriate termination mixes (*Table 4.9*) are added. Four microliters of the premixes are added to each of the four tubes per template and overlayered with 10 μl mineral oil.
8. The cyclic heat denaturation–sequencing process is initiated by heating to 95°C for 5 min before 30 cycles at 95°C/30 sec and 70°C/min.
9. The reaction is terminated with a sequencing gel loading buffer (95% (v/v) formamide, 10 mM EDTA (pH 7.6), 0.1% (w/v) xylene cyanol, 0.1% (w/v) bromophenol blue).
10. Gel electrophoresis: This is performed as described in Protocol 4.9 loading 2 μl/reaction of product on to the gel.

Table 4.9: PCR sequencing termination mixes

Termination mix	Constituents
A	15 μM dATP, dCTP, deaza-GTP, dTTP, 0.45 mM ddATP in $T_{10}E_{0.1}$
G	15 μM dATP, dCTP, deaza-GTP, dTTP, 0.03 mM ddGTP in $T_{10}E_{0.1}$
C	15 μM dATP, dCTP, deaza-GTP, dTTP, 0.3 mM ddCTP in $T_{10}E_{0.1}$
T	15 μM dATP, dCTP, deaza-GTP, dTTP, 0.9 mM ddTTP in $T_{10}E_{0.1}$

References

1. **Anderson, M.L.M. and Young B.D.** (1985) in *Nucleic Acid Hybridisation: a Practical Approach* (B.D. Hames and S. Higgins, eds). IRL Press, Oxford. p. 73.
2. **Li, Z. and Brow, D.A.** (1993) *Nucl. Acids Res.,* **21,** 4645.
3. **Lau, E.T., Kong, R.Y.C. and Cheah, S.E.** (1993) *Anal. Biochem.,* **209,** 360.
4. **Dean, M.** (1987) *Nucl. Acids Res.,* **15,** 6754.
5. **Leitch, A.R., Schwarzacher, T., Jackson, D. and Leitch, I.J.** (1994) In Situ *Hybridization.* BIOS Scientific Publishers, Oxford.
6. **Stahl, D.A., Krupp, G. and Stackebrandt, E.** (1989) in *Nucleic Acids Sequencing: a Practical Approach* (C.J. Howe and E.S. Ward, eds). IRL Press, Oxford. p. 137.
7. **De Wachter, R., Maniloff, J. and Fiers, W.** (1990) in *Gel Electrophoresis of Nucleic Acids: a Practical Approach* (D. Rickwood and B.D. Hames, eds). IRL Press, Oxford. p. 151.
8. **Gurr, S.J. and McPherson, M.J.** (1991) in *PCR: a Practical Approach* (M.J. McPherson, P. Quirke and G.R. Taylor, eds). IRL Press, Oxford. p. 147.

5 Quantitation of RNA

5.1 Spectrophotometric quantitation

One of the original methods for determining RNA concentration was the orcinol method [1]. This is a method which is quite sensitive down to less than 20 μg/ml. However, the reagent itself is quite hazardous and a number of compounds, especially sugars, interfere with this method. Spectrophotometric assessment of RNA concentration is now often based upon the fact that RNA absorbs light strongly at 260 nm. For this method it is very important that the RNA sample is not contaminated with phenol or any other compound that has an aromatic ring. The absorption is measured at the peak of the RNA spectra and the concentration can be calculated using the equation:

$$40 \times OD_{260} \text{ of the RNA sample} = \text{concentration of RNA sample } (\mu g/ml)$$

This equation is based on a sample of 40 μg/ml producing an OD_{260} of 1.0. The most accurate concentration for OD measurement is between an OD_{260} of 0.1 and 0.5. RNA samples should be diluted to a final volume of 0.5 ml and measured in quartz 0.5 ml cuvettes; be sure that there is no condensation on the outside of the cuvettes as a result of filling them with cold solution. Cuvettes should always be pretreated with 0.1% DEPC for 15 min at 37°C to inactivate ribonucleases. This method of quantitation is used in a preparative manner and not usually for analysis. The accuracy of this method does depend on the composition of the RNA, as all the bases absorb to different extents at 260 nm (*Table 5.1*). The advantage of this method over the orcinol method is that it is non-destructive in that the RNA can be used after it has been assayed. If the RNA is hydrolyzed then the previous equation

Table 5.1: UV absorption by nucleotides

Nucleotide	Molecular weight	Absorption maximum (nm)	Molar extinction in acid at 260 nm (10^3)	280/260
AMP	347	257	14.5	0.22
GMP	363	256	11.8	0.67
CMP	323	281	6.2	2.09
UMP	324	262	9.9	0.39

Molar extinction relates to single nucleotides not RNA.

is no longer valid and instead the molar extinction values given in *Table 5.1* apply.

The use of spectrophotometry can also provide a check upon the purity of the RNA preparation. Measurement at three separate wavelengths, 230, 260 and 280 nm, determines the ratio of polysaccharide:RNA:protein in a sample. Based on the average composition of RNA, the optimal ratio of optical density at 230, 260 and 280 nm is 1:2:1 which is usually judged to be pure enough for further preparative and analytical procedures.

Some spectrophotometers are not capable of scanning these wavelengths quickly, delaying measurements whilst the machine adjusts to the next wavelength to be measured. This has led to the production of spectrophotometric equipment designed purely for RNA or DNA quantitation (e.g. Genequant RNA/DNA calculator, Pharmacia). These DNA/RNA calculators are now becoming popular as their use can improve laboratory productivity.

Some RNA fractionation processes use UV scanning with quartz tubes to locate and sometimes quantitate the RNA species that have been separated. Polyacrylamide tube gels are often sectioned after locating bands to certain areas. Methods such as capillary electrophoresis (Section 5.6) also can be scanned at 260 nm to assess RNA location (w.r.t. fraction) and quantitate the peak area (RNA concentration) for each species.

5.2 Fluorescent quantitation

When there is insufficient RNA to be measured accurately by spectrophotometry it is possible to estimate the concentration of RNA by use of EtBr fluorescence. EtBr intercalates between nucleic acid bases and can then be excited by short-wavelength UV emitting visible light at 590 nm.

Protocol 5.1: Spot assessment of RNA concentration using EtBr

1. Spot 1–5 µl each RNA sample on to a sheet of Saran wrap (Dow Chemicals Company, available from GRI).
2. Spot equal volumes of RNA standards (0–20 µg/ml) which should be of a similar type of RNA as the unknown on to the Saran wrap.
3. To each sample and standard add an equal volume of EtBr (2 µg/ml in TE (10 mM Tris-HCl (pH 7.6), 1 mM EDTA)).
4. Mount the Saran wrap on a short-wavelength UV transilluminator and photograph the fluorescent spots. For photo-

graphing gels an orange filter is usually placed in front of the lens and typical settings of f4.5, 1/30–1/4 sec with Polaroid film are used (see Section 3.4.2).

5. RNA concentration is assessed by comparison of the fluorescence of the samples to the standards.

5.3 Autoradiographic quantitation

Quantitation of signal densities on autoradiographs or directly from radioactive blots is a technique of central importance in many molecular laboratories. The procedure used has been described in Section 3.5.2. The signals generated by radioactivity or chemiluminescence (Sections 3.5.2 and 3.5.3) are usually quantified by means of a laser densitometer. Measurement of the concentration of labeled RNA can only be achieved by correct use of laser densitometry equipment. The sensitivity of densitometers is sometimes lower than expected, where bands visible to the eye are sometimes not registered as above background and as such are not designated as bands by the computer.

Autoradiographs are scanned over a defined area of the densitometer grid system. Data are transferred to a computer software package which generally offers many useful options (see *Table 3.24*).

A fact of primary importance is the linearity of the response of the film, this can be achieved by preflashing the film [2]. Another important factor is the definition of a background density. This varies with exposure time, X-ray film type, film age and background radioactivity on the sample support (gel or membrane). Software packages can usually designate molecular sizes to bands by use of the mobility of the standards. They can also define volumetric density values for defined bands which is a valid measurement of RNA concentration. This is usually managed by outlining the band to be measured on a computer screen (which must not overlap with other bands) and selecting a volumetric integration of the area defined (*Figure 5.1*). This provides a better measure than assessment of lines drawn through the bands of interest and assessing peak areas.

Relation of the laser densitometric quantitation to an original RNA concentration must take into account any RNA losses in prior procedures. All the parameters of the preautoradiography methods should be known. Quantitation of RNA on autoradiographs, produced by radioactivity or chemiluminesence must account for RNA losses in blotting, probing and stripping procedures (Section 3.5.2). This can be assessed by carrying out these procedures of known concentrations of RNA alongside the original sample, measuring for any losses at each stage.

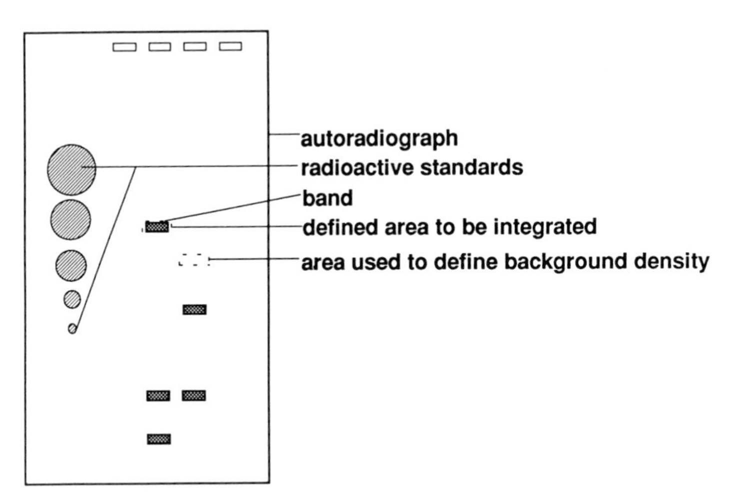

Figure 5.1: Volumetric density assessments using laser densitometry.

More recently, direct quantitation of radioactivity in gels and on membranes has become possible. Computerized systems such as the InstantImager™ (Canberra Packard) use microchannel array detectors (MICAD) and can image radioactivity generated by ^{14}C, ^{35}S, ^{32}P, ^{33}P and ^{125}I. The images form 100 times faster than images on X-ray film and they can be stored on computer.

5.4 Chemiluminescent-based measurements

As described in Section 3.5.3, systems that do not rely on exposure of X-ray film to light are becoming widespread. These systems directly assess the signal produced in a gel or on a membrane by recording light emissions. Signal emissions from chemiluminescent membrane-based systems (Section 3.5.3) can be directly recorded. Computerized systems linked to radiometric (photometer/luminometer) detection systems can capture images and quantify luminescence. Details of some commercially available instruments have been published [3], although recently released models will be in commercial catalogs. The basis of quantitation for these machines is an integration of the signal strength to yield a value which relates to the concentration of the RNA being probed.

Fluorescent primers are also commonly being used in sequencing protocols, replacing radioisotopic labeling methods in the larger laboratories. These systems have the distinct advantage of processing speed where immediate assessment of a sequence is possible instead of the possible 1–3 day delay necessary for exposing X-ray films.

5.5 Quantitative polymerase chain reaction (PCR)-based methods

The concentration of RNA species may be amplified by use of the PCR or non-PCR amplification procedures [4–7]. PCR is a revolutionary molecular technique capable of amplifying RNA species which are not detectable with sensitive techniques such as Northern analysis.

5.5.1 The polymerase chain reaction

The PCR was initially designed for amplification of DNA templates. A modification to the original procedure has enabled amplification of RNA by using the RNA as a template for cDNA synthesis with subsequent amplification of the cDNA by PCR.

PCR principles and reaction. PCR is based upon the copying of the two strands of a DNA template by a DNA polymerase, separating the strands produced and repeating the copying procedure with the larger number of templates available (*Figure 5.2*). This process involves the repetitive denaturation by heat which standard DNA polymerases used in other molecular biology procedures cannot withstand. A thermostable DNA polymerase (*Taq* polymerase) originally isolated from the thermophilic bacteria *Thermus aquaticus* can survive temperatures which are necessary to denature DNA templates and it is this and similar enzymes that are the cornerstones of PCR.

PCR reaction. PCR reactions can be carried out in 0.5 ml microcentrifuge tubes which fit into all commercially available thermal cyclers. These reactants include a template, primers, DNA polymerase buffer, dNTPs and DNA polymerase. Occasionally components such as glycerol (10%) and DMSO (5% w/v) are used to increase the efficiency and specificity of PCR reactions [7].

DNA template. The concentration of the DNA template can be quite variable according to the type of DNA to be amplified. Examples of template DNA can be a bacterial DNA, 50–500 ng genomic DNA, 5–10 μg sperm DNA or 1 μg RNA.

Primers. PCR primers are usually 16–35 bases in length. The choice and design of primers is often complex as their sequence leads to different types of behavior in the PCR reaction. Primer design must meet the following requirements.

(i) Primers should have no complementarity to each other as this would effectively leave few primers available to anneal to the template.

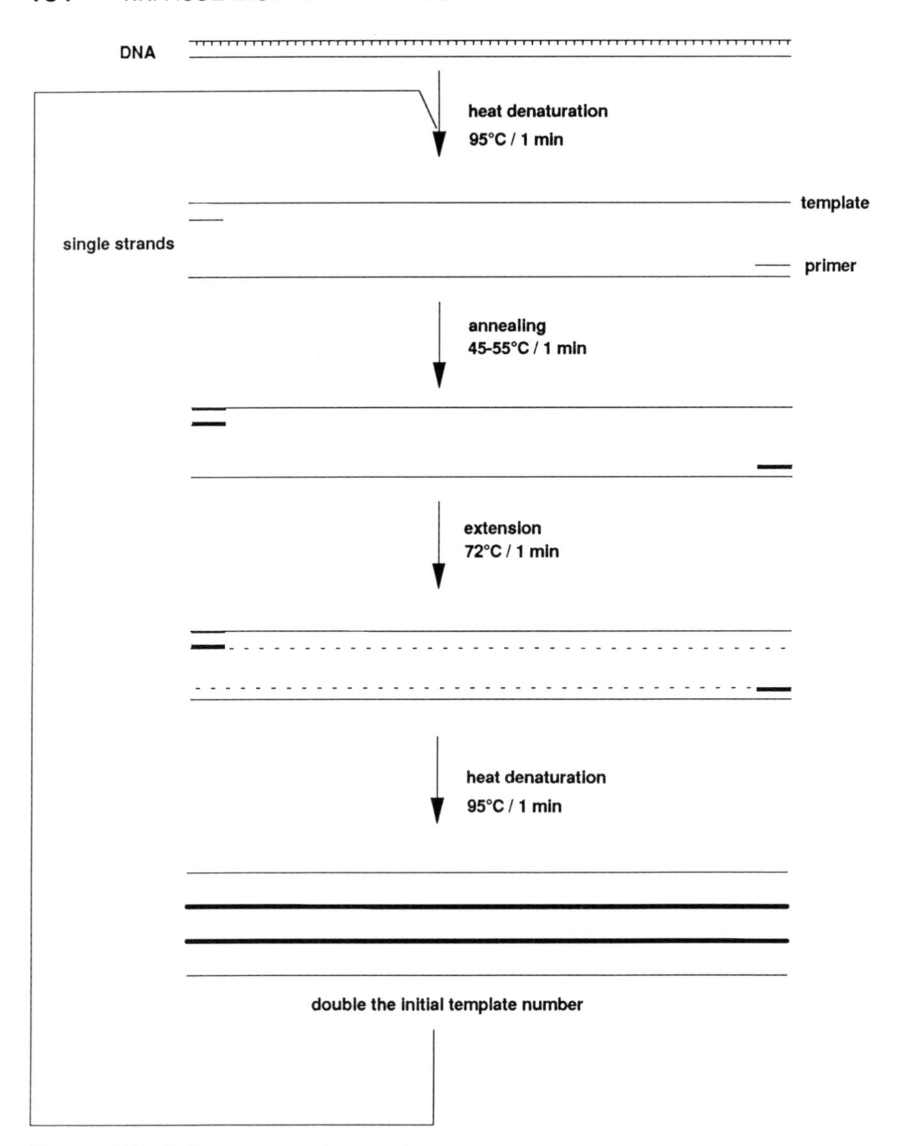

Figure 5.2: Polymerase chain reaction.

(ii) Primers should be unable to form secondary structures within their sequence (*Figure 5.3*).

(iii) Each primer pair should have a similar G + C composition so that annealing temperatures are similar.

(iv) The 3′ region of a primer must have absolute complementarity to the template.

(v) When special sequences (mutations or enzyme recognition sites) are to be included in a primer they should be positioned at the 5′ termini. Not surprisingly, computer software is now used extensively to help with primer design (e.g. Epicentre Software).

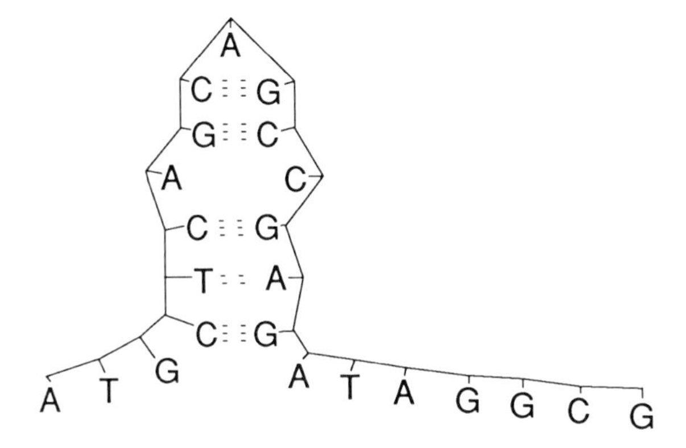

Figure 5.3: Secondary structure formation in PCR primers.

The concentration of primer in a PCR reaction can also be quite variable between different systems. Concentrations between 10 and 100 pmol (for each primer) have been used in PCR systems.

DNA polymerase buffer. Reaction buffers are supplied with the commercially available polymerases (*Table 5.2*). These are usually adequate for most amplifications although sometimes magnesium concentrations are varied to optimize amplification.

dNTPs. Deoxynucleoside triphosphates of the four bases are widely available (e.g. Pharmacia) and used at working concentrations of 0.5–5 mM.

DNA polymerase. The choice of DNA polymerase used to be restricted to *Taq* polymerase. Other polymerases are now available which are claimed to exhibit superior characteristics to *Taq* polymerase (*Table 5.3*). Appreciation of the importance of the heat stability of *Taq* polymerase at high temperatures led to the isolation of DNA polymerases with improved heat stabilities allowing amplification at higher temperatures consequently giving higher activities (*Figure 5.4*).

Table 5.2: DNA polymerase buffers

Polymerase	Source	Components
Taq polymerase	Cetus, Pharmacia, Boehringer Mannheim, IBI, Promega and others	100 mM Tris-HCl (pH 8.3), 15 mM $MgCl_2$, 500 mM KCl, 1 mg/ml gelatin (Boehringer Mannheim)
$Vent^R$ DNA polymerase, $Vent^R$ (exo^-) DNA polymerase, Deep $Vent^R$ DNA polymerase, Deep $Vent^R$ (exo^-) DNA polymerase	New England Biolabs	10 mM KCl, 20 mM Tris-HCl (pH 8.8 at 25°C), 10 mM $(NH_4)_2SO_4$, 2 mM $MgSO_4$, 0.1% (v/v) Triton X-100

Table 5.3: Thermostable DNA polymerases

Polymerase	Source(s)	Comments
Bacillus stereothermophilus DNA polymerase	Bio-Rad	Temperature optima of 65°C. Also used for sequencing
Taq polymerase	Cetus, Pharmacia, Boehringer Mannheim, IBI, Promega and others	Standard enzyme, prone to errors. 285×10^{-6} errors [8] per single round of gap filling, assessed by the opal reversion assay [9]
Thermus thermophilus DNA polymerase	USB	
Thermococcus litoralis DNA polymerase: Vent^R polymerase	New England Biolabs	Proofreading ability reduces errors. 57×10^{-6} errors [10]
Pyrococcus sp. DNA polymerase: Deep Vent^R	New England Biolabs	Proofreading ability eliminates errors

PCR cycles. The PCR reaction consists of three separate stages: template denaturation, primer annealing and chain extension. Each of these steps has a time and a temperature variable which must optimized for every system; this is most easily achieved using a thermocycler which can be programed in terms of the temperature and length of incubation for each step.

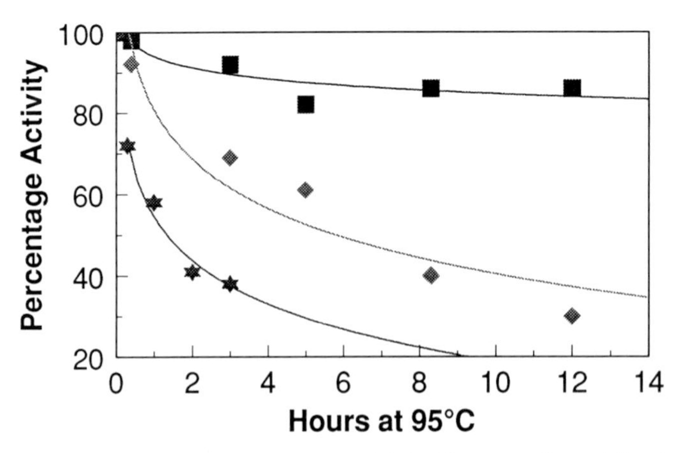

Figure 5.4: DNA polymerases used in PCR protocols: activity decline over time at 95°C. (■) Deep Vent DNA polymerase; (◈) Vent DNA polymerase; (★) *Taq* DNA polymerase.

LIVERPOOL JOHN MOORES UNIVERSITY
LEARNING SERVICES

Template denaturation. Templates are heat-denatured at the beginning of each PCR cycle. The temperature used to denature templates varies between 92 and 97°C. The temperature used is defined by the G:C contents and length of the template molecule. Templates with high concentrations of guanine or cytosine bases form stronger duplexes which require higher denaturation temperatures. Large templates also require higher denaturation temperatures as the melting temperature (T_m) of the template duplex is high. The use of high denaturation temperatures reduces the survival of *Taq* polymerase such that a balance must be found rather than opting for rapid template denaturation at high temperatures. Heat denaturation is often ensured prior to PCR by heating the reaction mixture to 95°C for 5 min and adding *Taq* polymerase during the first annealing stage. The length of time of the heat-denaturation steps is a function of template length. Large template molecules (1.5–5 kbp) take longer to denature making denaturation times of 1–2 min at 95°C necessary. Smaller templates (< 1.5 kbp) can be denatured by 30 sec at the appropriate temperature, usually 92–95°C.

Primer annealing. Most PCR primers are between 16 and 40 nucleotides in length and can be exact matches of template sequence or can be a mixture of primers where one or more of the positions are random nucleotides (degenerate primers). Degenerate primers are used when the template sequence is not absolutely defined (e.g. heterologous systems) or when multiple targets are to be amplified.

The degree of matching between the primer and template is the major factor determining annealing temperature. Primers showing a perfect match can be annealed with high stringency 55–72°C whereas degenerate primers should be annealed at lower stringencies (45–55°C) to allow annealing to occur. The other factor determining the annealing temperature is the G:C content of the primer. As a guide, an initial annealing temperature for hybrid formation can be calculated on the basis of 2°C for A:T/U pairings and 3°C for G:C pairings. For example, the primer GGCCATTCACCACCTTTGGGGCCCC can initially be annealed to its template at 67°C (17 (G:C) × 3°C + 8 (A:T) × 2°C).

Annealing stages last between 30 and 90 sec according to the stringency of the match between primer and template; the shorter the time the higher the stringency.

Chain extension. Chain extension is normally carried out at 72°C. The extension time period is defined by the length of the region to be copied. As a guideline, the maximum rate of *Taq* polymerase can be assumed to be 200 bases/sec.

LIVERPOOL JOHN MOORES UNIVERSITY
LEARNING SERVICES

Cycle number. The number of cycles in a given procedure is normally between 20 and 60. Amplifications from concentrated samples can be achieved with a low number of cycles whereas rare species often require more cycles and sometimes a fresh addition of *Taq* polymerase to ensure that the reaction continues. The PCR process usually gives exponential amplification through the early cycles which flattens out into a plateau (*Figure 5.5*) at the longer times. The plateau effect is caused by a limitation of substrates (dNTPs and primers) and reduced enzyme activity. Samples that have been amplified into this plateau range are difficult to quantify.

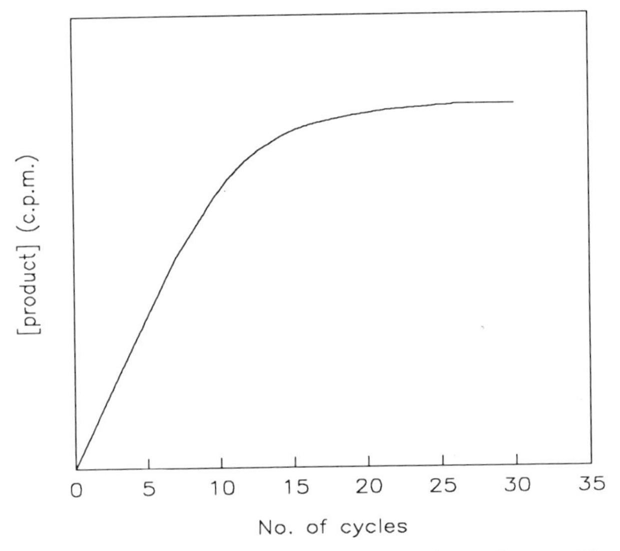

Figure 5.5: PCR product plotted over time: plateau effect.

Samples are usually qualitatively assessed by agarose gel electrophoresis to ascertain whether amplification has occured. Visualization of amplified products can be achieved using ethidium bromide, fluorescent tagged primers or with colorimetric detection assays [11].

The PCR technique, though powerful, is sometimes temperamental. Certain parameters, especially primer design, [Mg^{2+}], annealing temperature and extension time, are required to be defined in many circumstances.

5.5.2 cDNA synthesis and amplification of cDNA by PCR

The basis of PCR quantitation of RNA involves the synthesis of a cDNA molecule using the RNA as a template. The cDNA can then be used as a template for PCR [12–16].

Protocol 5.2: cDNA synthesis using reverse transcriptase

1. Reaction mixture:
   10 µg total RNA
   50 pmol primer (complementary to the RNA sequence)
   1 µl 10 mmol each dNTP (BRL)
   1 µl (0.2 U) reverse transcriptase (BRL)
   1 µl (10 U) RNasin (RNase inhibitor)
   to a final volume of 10 µl
2. Reverse transcription is initiated by transfer to the thermal cycler and incubating for 30 min at 42°C.

Protocol 5.3: PCR amplification

1. Before amplification the reverse transcriptase enzyme must be inactivated to avoid two copying processes occuring at once. This is achieved by heat-inactivating the enzyme for 5 min at 94°C.
2. PCR reaction:
   4 µl *Taq* polymerase buffer (100 mM Tris-HCl (pH 8.8), 500 mM KCl, 0.1% (w/v) gelatin, 1% (v/v) Triton X-100, 1.5 mM $MgCl_2$)
   50 pmol primer complementary to the cDNA strand
   34.5 µl sterile, double-distilled water
   0.5 µl *Taq* polymerase (Pharmacia).
3. The reaction mixture in the tube is overlayered with 100 µl mineral oil to prevent loss by evaporation.
4. The cycle parameters used are:
   template denaturation–94°C for 40 sec
   primer annealing–55°C for 45 sec
   chain extension–72°C for 1 min
   number of cycles–40.

Standard PCR reactions have an inherent variability in the amount of amplification that is achieved with low concentrations of template and this makes accurate quantitation difficult. At high concentrations of template, amplification is limited by the amounts of substrates (dNTPs, primers and magnesium ions) allowing accurate quantitation. Template levels produced from reverse transcription are rarely highly concentrated, making quantitative PCR more complex.

5.5.3 Quantitative PCR using internal controls

Amplification of cDNA by PCR has inherent variability. This can be bypassed with varying degrees of efficiency. Amplification using an internal control as a test of the system can be used to allow quantitation to be carried out [17] as follows.

LIVERPOOL
JOHN MOORES UNIVERSITY
AVRIL ROBARTS LRC
TEL. 0151 231 4022

Protocol 5.4: Quantitative PCR with co-amplification of control cDNAs

1. In this method mRNA isolated using an affinity chromatography protocol (Section 2.9) is used as a template for reverse transcriptase.
2. The cDNA is then amplified using the cycle characteristics as described previously or 95°C for 1 min, 54°C for 1 min, 72°C for 2 min (40 cycles) and 7 min at 72°C to ensure complete strand synthesis before analysis. Internal controls (primers to other genes) can be used with the same mRNA samples to amplify other known genes as confirmation that PCR is proceeding correctly.

Analysis of the products can be achieved by measuring optical density [18].

5.5.4 Amplification with reference to external controls

This procedure characterizes the system fully before amplification of an unknown quantity of RNA by repeated amplification from known amounts of targets over time. This allows reliable interpretation of a given amplification when amplifying from sample containing an unknown amount of template [19]. Repeated amplification of known concentrations of cDNA targets, incorporating ^{32}P-labeled dCTP for 20, 30 and 35 cycles allows construction of an amplification plot by averaging the concentration of amplified product at each time point for each template concentration. This plot will define the variability of the process and allow interpretation of results from sample amplifications.

PCR reactions include 37 kBq (1 μCi) of [^{32}P]dCTP with cycle characteristics appropriate to the system. Samples are then electrophoresed and bands (visualized by EtBr staining) are excised. Quantitation is carried out in a scintillation counter.

Plots of radioactivity incorporated against cycle number allow linear relationships to be generated (*Figure 5.5*) for any given target.

5.5.5 Kinetic PCR analysis

Kinetic analysis of PCR reactions allow quantitation of the RNA by capturing data from the exponential phase of a PCR reaction [20]. This method uses ethidium bromide which fluoresces when intercalated with nucleic acids. As PCR proceeds fluorescence is monitored by a video camera during each annealing stage. The video image is 'grabbed' by computer for later quantitation. EtBr at a concentration of 4 μg/ml is used in standard 100 μl reactions under conditions suitable to the system. The camera apparatus is fitted with a 600 nm interference filter to ensure fluorescence of EtBr is efficiently captured. UV illuminators are set up behind the camera to initiate fluorescence in the reactions.

5.5.6 Competitive PCR

Competitive PCR bypasses the variability of amplification of low-level templates by coamplifying two templates using the same primers and assessing the ratio of the products [21]. This method can be used on well-characterized mRNA extracted from tissue using one of the techniques described in Chapter 2. Well-characterized mRNA is the only valid target for this procedure as transcripts for duplicated genes would all be amplified as would processed RNA. Therefore, mRNA characterized by Northern blotting, probing for heterogeneity and nuclease protection assays (Section 4.3.4) can be assessed using this method. Reactions are carried out with two templates which are the cDNA of interest and a DNA fragment. The DNA fragment must have identical sequence to the RNA at the positions at which the primers bind. DNA templates that fulfil these requirements are genomic fragments which encode the cDNA or random DNA that has had primer sites engineered (by cloning or PCR) into it. The latter option is sometimes the only possibility as genomic DNA is not always available and, as primer sequence is usually known, engineering DNA is relatively easy.

The main requirement is that the DNA and RNA (cDNA) can be differentiated. One method for achieving this is to vary the size of the RNA and DNA by at least 100 bases, assessing the products by end-labeling the primers (Section 3.5.1) before amplification and Northern blotting. With genomic DNA templates this is possible by use of a region that encodes an intron. A difference in size between the two templates may cause a variability in the reaction, making assessment of the ratio ineffective for quantitation. Differentiation by using a synthetic DNA template of equal size but encoding a restriction site not found in the cDNA would allow differentiation by enzymatic digestion of the amplified products (and subsequent Northern blotting). The amplification procedure coamplifies known amounts of DNA with set amounts of RNA isolate. In this manner the ratio of concentration of a single RNA species may be assessed by autoradiography and laser densitometry of the amplified sequence. An alternative to a synthetic DNA target is use of the RNA species under study that has been mutagenized to produce a restriction site not possessed by the original RNA. This method has been dubbed PCR aided transcript titration assay (PATTY) [22].

5.6 Free solution capillary electrophoresis

As described in Section 3.3, electrophoresis is a valuable tool for characterizing RNA. However, for quantitating RNA, capillary electrophoresis, electrophoretic migration through long, thin tubing (60 cm × 100 μm diameter), is an extremely useful tool. A number of manufacturers make capillary electrophoresis apparatus suitable for RNA separation and quantitation. Standard buffers such as Tris-borate can be used to provide the solution through which charge can flow. Samples can be applied under vacuum to achieve tight band formation. Electrophoretic settings vary with different systems and with different requirements.

RNA applied to capillary electrophoretic procedures should be a single or just a few species to avoid spectral overlap.

Protocol 5.5: Capillary electrophoresis of a biotinylated mRNA species

1. Purification of specific RNA species is possible by capturing with biotinylated probes hybridized to avidin or to magnetic beads (Section 2.9.2) [17]. The amount of total RNA used in the affinity capture is 20–200 μg.
2. After capture the samples are concentrated by ethanol precipitation and redissolved after pelleting in 5 μl sterile water.
3. Free solution capillary electrophoresis linked to a UV detector (Prime Vision, system IV, Europhor, France) is then used to analyze the sample.
4. A 100 μm capillary, 60 cm long with the detector positioned 45 cm from the loading position has the sample loaded under vacuum for 0.5 sec.
5. Electrophoresis with a TBE buffer (89 mM Tris, 89 mM boric acid, 2 mM EDTA, pH 8.3) at 400 V/cm for 10–30 min separates the species involved and the detector quantifies the species passing it.

References

1. **Schneider, W.C.** (1957) *Meth. Enzymol.*, **3**, 680.
2. **Lasky, R.A.** (1989) in *Radioisotopes in Biology: a Practical Approach* (R.J. Slater, ed.). IRL Press, Oxford. p. 87.
3. **Walker, G.T., Fraiser, M.S., Schram, J.L., Little, M.C., Nadeau, J.G. and Malinowski, D.P.** (1992) *Nucl. Acids Res.*, **20**, 1691.
4. **Kwoh, D.Y., Davis, G.R., Whitfield, K.M., Chappelle, H.L., Dimechele, L.J. and Gingeras, T.R.** (1989) *Proc. Natl Acad. Sci. USA*, **86**, 1173.
5. **Kuppuswamy, M.N., Hoffmann, J.W., Kasper, C.K., Spitzer, S.G., Groce, S.L. and Bajaj, S.P.** (1991) *Proc. Natl Acad. Sci. USA*, **88**, 1143.

6. **Guatelli, J.C., Whitfield, K.M., Kwoh, D.Y., Barringer, K.J., Richman, D.D. and Gingeras, T.R.** (1990) *Proc. Natl Acad. Sci. USA,* **87**, 1874.
7. **Lu, Y-H. and Negre, S.** (1993) *Trends Genet.,* **9**, 297.
8. **Tindall, K.R. and Kunkel, T.A.** (1988) *Biochemistry* **27**, 6008.
9. **Kunkel, T.A.** (1987) *Proc. Natl Acad. Sci. USA,* **84**, 4865.
10. **Mattila, P., Ronka, J., Tenkanen, T. and Pitkanen, K.** (1991) *Nucl. Acids Res.,* **19**, 4967.
11. **Kemp, D.J.** (1992) *Meth.Enzymol.,* **216**, 116.
12. **Conrad, R., Liou, R.F. and Blumenthal, T.** (1993) *Nucl. Acids Res.,* **21**, 913.
13. **Don, R.H., Cox, P.T. and Mattick, J.S.** (1993) *Nucl.Acids Res.,* **21**, 783.
14. **Lambert, K.N. and Williamson, V.M.** (1993) *Nucl. Acids Res.,* **21**, 775.
15. **Lozano, M.E., Grau, O. and Romanowski, V.** (1993) *Trends Genet.,* **9**, 296.
16. **Goblet, C., Prost, E., Bockhold, K.J. and Whalen, R.G.** (1992) *Meth. Enzymol.,* **216**, 160.
17. **Reyes-Engel, A. and Dieguez-Lucena, J.L.** (1993) *Nucl. Acids Res.,* **21**, 759.
18. **Wang, A.M., Doyle, M.V. and Mark, D.F.** (1989) *Proc. Natl Acad. Sci. USA,* **86**, 9717.
19. **Irving, J.M., Chang, L.W.S. and Castillo, F.J.** (1993) *Bio/technology,* **11**, 1042.
20. **Higuchi, R., Fockler, C., Dollinger, G. and Watson,R.** (1993) *Bio/technology,* **11**, 1026.
21. **Gilliland, G., Perrin, S., Blanchard, K. and Bunn, H.F.** (1990) *Proc. Natl Acad. Sci. USA,* **87**, 2725.
22. **Becker-Andri, M. and Hahlbrock, K.** (1989) *Nucl. Acids Res.,* **17**, 9437.

6 Functional analysis of RNA

6.1 Basic principles

The ultimate test for any purified RNA is whether it functions *in vitro* as it does *in vivo*. In the case of mRNA, the simple test is whether it directs the correct synthesis of a polypeptide chain as it does in the intact cell. However, a significant proportion of RNAs do not code for protein and so other criteria must be used to judge their functionality. It has become apparent that secondary structure and the ability to bind proteins in a highly specific manner are extremely important for the function of many types of RNA such as rRNA, tRNA and the snRNAs involved in splicing reactions. In addition to RNA–protein interactions, RNA–RNA interactions have also assumed greater importance with the discovery and characterization of RNA catalysis. Hence, when elucidating the function(s) of RNA it is necessary to use a variety of techniques. It is not possible within the confines of a single chapter to give detailed protocols for all the techniques used in analyzing the functional aspects of RNA. For this reason the authors have chosen to give detailed descriptions of the most widely used techniques and provide an overview of the other methods that are available.

6.2 Computer analyses of sequences

With the advent of rapid sequencing methods (Section 4.4), an increasing number of RNA sequences have become available and these have enabled extensive analysis of RNA by a variety of computer techniques [1]. In the case of mRNA it is possible to recognize not only the open reading frame and the primary sequence of the protein that it codes for, but also the motifs both within the coding region and in the non-translated parts of the mRNA. Of equal importance is the ability to model the higher order structures of all types of RNA since it has become clear that many of the functional properties of RNA are determined by the higher order structure of the molecules; this is true not only of tRNA and rRNA but also mRNA. Most of these analyses use one of the commercial software packages most of which can be run on IBM MSDOS personal computers or Macintosh systems possessing powerful processors. Software programs offer many facilities that are tedious or impossible to accomplish by eye. These facilities include the following.

(i) Sequence homology scanning, for example between a newly sequenced RNA and any other RNA sequences in storage (most comparisons are performed on network systems with massive database storage facilities).
(ii) RNA sequences are often limited to short regions of reliable sequence which must be aligned with each other at overlaps to generate the complete sequence. Alignment is also commonly used to assess homologies in a set of RNA sequences. This type of study can reveal sequence differences that may have an eventual phenotype, function or pathogenicity.
(iii) Structural modeling is still not a perfected science on computer systems as high-level analysis systems (X-ray crystallography and nuclear magnetic resonance (NMR)) are necessary for complete definition of structure. Computer packages can, however, predict possible base pairing which is the secondary structure that plays such a vital role in RNA function. The most common use of this function is the prediction of loop structures in RNA.
(iv) mRNA has characteristic sequences that delineate the coding and non-coding regions. These can be matched to an RNA sequence to suggest where the coding regions begin and end.

Table 6.1 lists some of the packages available for analyzing nucleic acid sequences. In addition, there are major international databases of sequences available which will allow comparison of the sequence of RNA to known sequences; this can be invaluable for looking for sequence homologies and can often indicate the most promising direction of research. Most commercially available software packages for personal computers are available in IBM-compatible and Macintosh formats due to the different preferences of research institutes.

Table 6.1: Software packages for nucleic acid analysis

Software package	Operating system	Source
DNASTAR	MSDOS	Dnastar Inc.
DNASIS	MSDOS	Hitachi Software
IBI/Pustell	MSDOS	IBI
Microgenie	MSDOS	Beckman
UWCGG	UNIX	University of Wisconsin
Staden programs	VAX	Amersham

6.3 Analysis of proteins binding to RNA

6.3.1 Introduction

The binding of proteins to RNA has been widely studied both as a measure of the integrity of RNA and for studying the different functions of various proteins attached to RNA. In early work complete reconstitution of nucleoproteins such as bacterial ribosomes helped to elucidate the role of some ribosomal proteins. However, such reconstitution procedures are often difficult to carry out given the complexity of not only protein–RNA interactions but also protein–protein interactions.

Four main methods are used to study the interactions of protein and RNA which have the same incubation procedures but differ in separation methodology to remove unbound RNA from bound RNA. These methods are filter assay, gel retardation analysis, sucrose gradient centrifugation and affinity chromatography.

6.3.2 Formation of protein–RNA complexes *in vitro*

Incubation conditions. Assays of protein interactions with RNA are carried out under optimum conditions in terms of the ionic content of the assay solutions that favor specific binding. Variables can include pH, ionic strength, divalent cations and an energy source such as ATP. *Table 6.2* gives details of some of the solutions used for binding studies.

Binding protocols are usually optimized for individual systems by testing for the optimum binding conditions using known samples.

One important experimental factor, when studying protein–nucleic acid interactions is to prove that the interaction is of a specific nature. This is particularly important because many proteins will bind

Table 6.2: Examples of incubation conditions used for studying *in vitro* RNA–protein interactions

Binding solution	Constituents	Example system	Example incubation conditions	Ref.
1	40 mM Tris-HCl (pH 7.9), 50 mM KCl, 1 mM DTT, 2 mM PMSF, 100 μg yeast tRNA, 4 U RNasin (Promega)	Recognition of small RNA molecules by HIV-1 Rev protein	4°C for 20 min	2
2	40 mM Tris-HCl (pH 7.4), 40 mM KCl, 5% w/v) glycerol and 0.01% bromophenol blue	Interaction of an iron-responsive element and its cognate RNA	Heat-denatured RNA (65°C for 5 min with snap cooling) is mixed with protein sample and loaded on to gel	3
3	10 mM Tris-HCl (pH 7.4), 2.5 mM $MgCl_2$, 0.5% (v/v) Triton X-100	Product of the fragile X gene binding to its cognate RNA	Rocking for 10 min at 4°C	4
4	20 mM Tris-HCl (pH 7.5), 100 mM KCl, 0.1 mM EDTA, 5 mM 2-mercaptoethanol	Affinity labeling of 40S ribosomal subunits	10 min at 37°C	5

non-specifically to nucleic acids. Specific binding can be achieved by carrying out the binding assay in the presence of appropriate competitor molecules. Competition assays are based on the fact that radiolabeled species (protein or RNA) can undergo equilibrium binding which will allow a protein to find the RNA sequence with which it binds specifically even when a large excess of non-specific competitor molecules are added. Competitor molecules can be specific or non-specific; representing the normal positive and negative controls of a binding experiment, respectively.

Specific competitors would be unlabeled nucleic acid of identical sequence to the species under study or unlabeled protein identical to the protein species under study. A specific competitor when added in excess to a binding reaction should displace the labeled portion of the complex. This occurs as the binding is in constant molecular equilibrium where molecules are releasing and rebinding over the time of the incubation.

Non-specific competitors are often random pieces of nucleic acid or polyribonucleotides. These should not contain the RNA sequence(s) that are able to compete for binding in a specific complex. Non-specific RNA is often included in all binding reactions (e.g. 40 μg yeast tRNA (Gibco–BRL)) to prevent non-specific binding in much the same way as is used to prevent non-specific binding to membranes for Northern blotting.

The amounts of RNA and protein used in a binding reaction depend on the availability of the material and the labeling method. When using low concentrations of RNA or proteins then it is necessary to use a high-activity label (^{32}P-phosphorus). *Table 6.3* details the amounts of RNA and protein used in selected binding reactions.

Cross-linking protein–RNA complexes. The protein–RNA complexes formed in incubations described previously may be converted into permanent complexes by cross-linking the molecules. Cross-linking is carried out in order to allow procedures that would otherwise disrupt the complex or in which fixed binding is essential. RNA binding site studies and RNA binding site mutagenesis studies are examples where specific binding must be preserved during analysis. This can be achieved using UV or formaldehyde cross-linking.

(i) *Ultraviolet cross-linking.* This method is the most commonly used cross-linking method. Specific apparatus (e.g. Ultraviolet Products model UVG-54) are used to supply a known dose of irradiation to the sample. The RNA–protein complex is exposed to intensities of 1.8 mW/m^2 for 5 min at 30°C. The samples may then be used in further methodologies. It should be noted, however, that this process is not reversible.

Table 6.3: Target molecule concentrations in selected binding assays

Example system	RNA concentration	Protein concentration	Ref.
Recognition of small RNA molecules by HIV-1 Rev protein	5 fmol ^{32}P-labeled RNA	1.3 ng Rev protein	2
Interaction of an iron-responsive element and its cognate RNA	1 ng ^{32}P-labeled RNA	3.2 μg total protein from a cell lysate	3
Product of the fragile X gene binding to its cognate RNA	20–37.5 μg homopolymer RNA bound to Sepharose beads	100 000 c.p.m. *in vitro* produced protein [^{35}S]methionine	4
Affinity labeling of 40S ribosomal subunits	3.3 pmol Met-tRNA^{met} and 100 pmol ribo-oligonucleotide	10 pmol 40S subunits	5

(ii) *Formaldehyde cross-linking.* Protein–RNA complexes can be cross-linked by incubating extracts in 1% (w/v) formaldehyde in a HEPES buffer (neutral pH) for 24 h at 0°C. After fixation, the formaldehyde must be removed by dialysis in dilute buffer. This type of fixation can be reversed if required by heating the samples for prolonged periods of time [6,7].

6.3.3 Separation of protein–RNA complexes

Filter/membrane assay. The nitrocellulose filter or membrane assay was one of the earliest methods developed to study protein–nucleic acid interactions and it is still a rapid and widely used method. The protein is incubated with radiolabeled RNA and then rapidly filtered under suction through a nitrocellulose membrane. In an ideal case, the RNA binds tightly to the membrane, and any protein bound to the RNA is also retained. The radioactive complexes can then be detected by scintillation counting. Various concentrations of the protein can be incubated with a fixed amount of RNA to assess binding affinities. This method has been widely used for many rRNA studies.

However, because the separation depends on the affinity of proteins for nitrocellulose disks, it does not always work; for example, complexes of ribosomal proteins with large rRNAs are not retained by the filter. It should be emphasized that this filter assay is a non-equilibrium method when reaction times are short as usually equilibrium can only be reached by extended incubation times. If the reaction is permited to reach equilibrium then rapid immobilization by suction of the sample through the nitrocellulose disk will capture the reaction at equilibrium point. Binding constants have been obtained by this method for a number of RNA–protein complexes [8]. Binding constants are derived using complex equations based on data from experiments which capture complexes at equilibrium.

Gel retardation analysis/gel shift assays. Gel retardation analyses (gel shift assays) are based on the observation that the mobility of a specific RNA–protein complex in a polyacrylamide gel is much less than that of the free RNA. The principles of this technique are illustrated in *Figure 6.1.* This procedure allows the separation and quantification of the free and complexed components of a binding reaction. The degree of retardation of the complex can be enhanced by adding antibodies to the proteins of the protein–RNA complex thus significantly increasing the size of the complex and so reducing its mobility. In addition, complexes involving multiple proteins or multiple classes of proteins can be separated, which allows the method to be applied to solutions containing a mixture of binding activities. Finally, it is possible to isolate the complex from the gel and characterize it further (e.g. Section 3.3). Alternative terms used to describe this type of method are 'gel shift' or 'gel mobility' analysis.

The proteins and RNA of interest are mixed together with competitor RNA (Section 6.3.2) which ensures that proteins do not bind non-specifically to the target RNA. As stated previously, the binding specificity is very dependent on the ionic conditions. In some cases, it may require other factors such as ATP, depending on the nature of the protein and RNA. The solution containing the mixture of protein–RNA complexes and unbound molecules is subjected to non-denaturing polyacrylamide gel electrophoresis (Section 3.3). Gel concentrations are defined by the size of the complex, although a 6.5% polyacrylamide (cross-linking ratio of 60:1), 0.5% agarose, 1 × TG (100 mM Tris-HCl, pH 8.8, 50 mM glycine) is a good starting point for new systems.

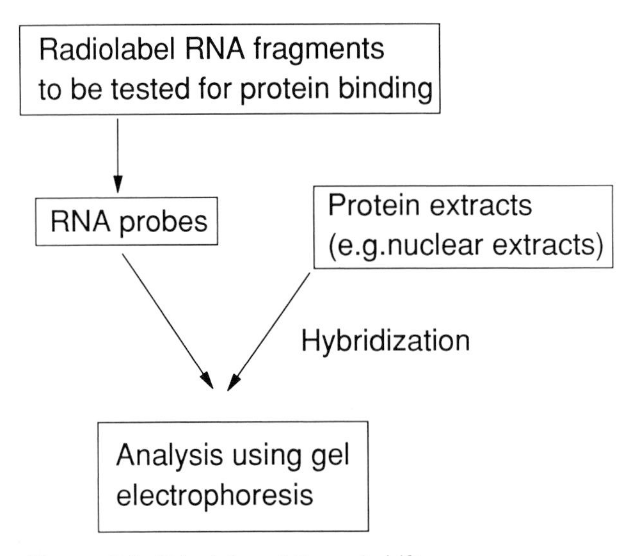

Figure 6.1: Principles of the gel shift assay.

The highly negatively charged, free RNA will rapidly enter the gel as a band, physically removed from the other components whilst the highly negatively charged RNA–protein complexes move into the gel, but are retarded, moving more slowly through the gel matrix than the free RNA. The amount of uncomplexed RNA does not affect any molecular equilibration, either in the binding reaction, whilst being loaded on to the gel, or during electrophoresis. Any RNA released by dissociation during the run will trail behind the free RNA band as a faint smear, and is usually not detectable. Hence, the assay yields an accurate value for the amount of uncomplexed RNA in the original solution, provided that no significant association or dissociation occurs in the 'dead time' between when the solution is loaded on to the gel and when the free RNA enters the gel. In practice, this is not a serious limitation, since specific RNA–protein complexes are usually quite stable. From the concentration of free RNA and the known amount of RNA in the binding assay, the amount of RNA complexed to protein can be calculated. Furthermore, in many cases the complexes themselves are stable during electrophoresis, and hence can be quantified, or studied for their stoichiometry.

After electrophoresis it is necessary to be able to detect the free and complexed RNA separate from the competitor RNA and its non-specific complexes. The most convenient way to do this is to radiolabel the RNA of interest with ^{32}P-phosphorus (Section 3.5.1) and then the position of the free and complexed RNA in the gel can be detected using autoradiography. An example of this type of analysis is shown in *Figure 6.2.* If for some reason it is not convenient to label the RNA, the position of the free RNA and RNA–protein complexes in the gel can be determined by blotting on to a membrane and then detecting the position of the RNA by hybridization with a ^{32}P-labeled probe complementary to the RNA of interest (Section 3.5.2).

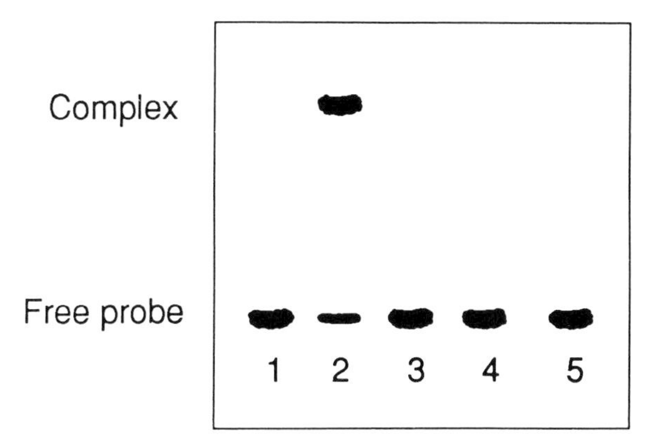

Figure 6.2: Analysis of gel shift assay using an autoradiograph. (1) Proble only; (2) probe + RNA 1 – complex formed; (3) probe + RNA 2; (4) probe + RNA 3; (5) probe + RNA 4.

Size separations using rate-zonal gradient centrifugation and gel permeation chromatography. Rate-zonal sedimentation and gel permeation chromatography have also been used to study the binding of ribosomal proteins to rRNA. These methods have advantages over electrophoresis in that a wider range of salt concentrations can be used, and the RNA or protein concentrations can be more readily estimated. Rate-zonal sedimentation can be used as a binding assay only when the protein–RNA complex has a significantly greater sedimentation rate than the protein alone. This limits its application to ribosomal proteins binding rRNA fragments larger than 5S. However, once the binding constant for a large RNA is determined, the relative affinity of small fragments may be found using a competition type of experiment. In the case of gel permeation chromatography, some preferential losses of complexes may occur on the column and, as in the case of centrifugation, this method works best with more stable complexes.

Affinity chromatography using biotinylated RNA. Affinity chromatography can be used to study the interactions of proteins with RNA. *Figure 6.3* is a schematic diagram showing how affinity chromatography can be used to study the interaction between protein and RNA in spliceosomes [9]. snRNA molecules bind to pre-mRNA in splicing reactions. This can be achieved *in vitro* by synthesis of the RNA molecules using *in vitro* transcription systems. The pre-mRNA is synthesized *in vitro* (Section 3.5.1) and is hybridized in solution to a small biotinylated 'anchor' RNA complementary to either the 3′ end or the 5′ end of the RNA. The RNA hybrid is immobilized on a streptavidin-coated magnetic bead. The protein extract is then added to the immobilized RNA, and during the incubation some of the RNA is spliced. The immobilized RNA is washed and the RNA and any components bound to it are eluted by reducing the disulfide bonds of the bio-dUTP nucleotide analogs of the anchor RNA. The resultant RNA molecules are analyzed with denaturing gel electrophoresis [9].

6.3.4 Determination of binding sites and protein species essential to protein–RNA reactions

Basic principles. Two main avenues are usually explored when a complex interaction is found. Firstly, the binding site of a RNA molecule to a protein is of interest and secondly the necessity of the various elements of the sometimes highly complex reactions (e.g. splicing).

Binding sites are usually defined with respect to RNA sequence. Binding sites can be located to a region of RNA by modifying the RNA at specific regions and ascertaining whether the protein–RNA complex is still formed. For example, modification of a pre-mRNA may

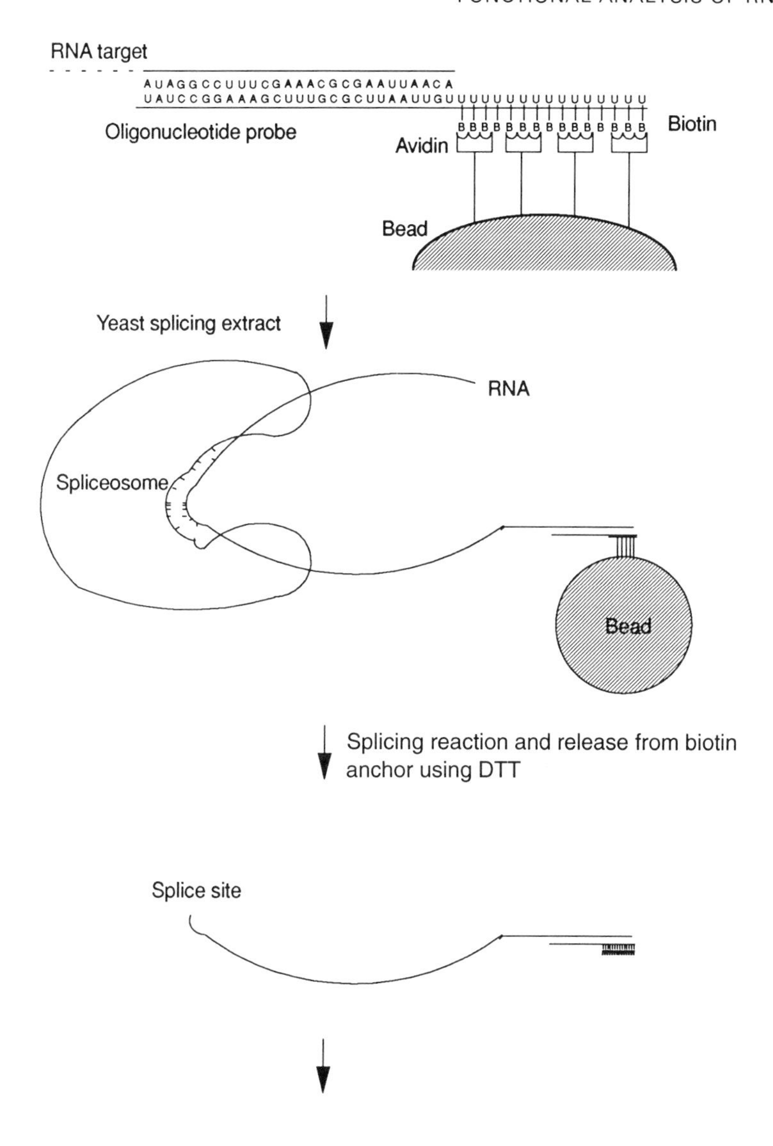

Figure 6.3: Affinity chromatography with reversably biotinylated RNAs: solid phase splicing.

inhibit spliceosome formation *in vitro* so that the RNA–protein complex is not seen on a polyacrylamide gel. This method of analysis is called modification–interference analysis and is usually used for spliceosome studies.

When binding sites have been located, mutagenesis of *in vitro* transcribed RNA can be used to define essential RNA bases testing for binding with various point mutants. This is achieved by cross-linking the protein–RNA complexes and digesting the unprotected RNA with ribonucleases such as RNase T1.

The following paragraphs describe in detail the use of modification–interference analysis to study the binding of proteins to RNA and provide an overview of some of the other methods used for studying the binding of proteins to RNA.

Determination of binding sites by modification–interference analysis [10]. This method uses RNA molecules synthesized *in vitro* to study protein binding sites. The RNA molecules are end labeled (Section 3.5.1) with ^{32}P. Chemical modification is usually achieved using diethyl pyrocarbonate (DEPC) or hydrazine. These chemicals will prevent formation of RNA–protein complexes. DEPC carboxyethylates purine bases and hydrazine removes pyrimidines. The procedures aim to modify only one base per RNA molecule. The action of both agents provides points of attack for aniline. These cleaved molecules will not be able to bind to proteins. The binding site is defined by autoradiographs of radiolabeled fragments on polyacrylamide gels where fragments that could have bound proteins (if they had not been modified and cleaved) do not appear (*Figure 6.4*).

Protocol 6.1: Modification–interference analysis

The minimum amount of RNA needed for this analysis is 40 fmol RNA (10^6 c.p.m.). The RNA can be synthesized *in vitro* or isolated by specific biotinylation and affinity chromatography or immunopurification. The RNA is then end labeled and purified by PAGE. The RNA is eluted and aliquoted into the number of reactions required. Non-specific RNA (e.g. yeast tRNA) is added to a final concentration of 12.5 μg. The RNA is ethanol precipitated and rinsed with 70% ethanol before being dried.

1. Purine modification. RNA pellet is redissolved in 200 μl purine modification buffer (50 mM sodium acetate (pH 4.5), 1 mM EDTA) and 2 μl fresh DEPC is added. The sample is mixed by vortexing and modification proceeds at 95°C for 2 min. The reaction is terminated by precipitating the RNA (75 μl 1 M sodium acetate (pH 4.5), 750 μl ethanol). RNA is pelleted by centrifugation at 4°C for 20 min. RNA is redissolved in 200 μl 0.3 M sodium acetate (pH 3.8) and 600 μl ethanol is added. The precipitated RNA is collected by centrifugation as before.
2. Pyrimidine modification. This can modify the bases specifically (C or U) or modify both kinds of pyrimidines.
   (i) To modify C residues only, the RNA is redissolved in 20 μl fresh anhydrous hydrazine, 3 M NaCl. Modification

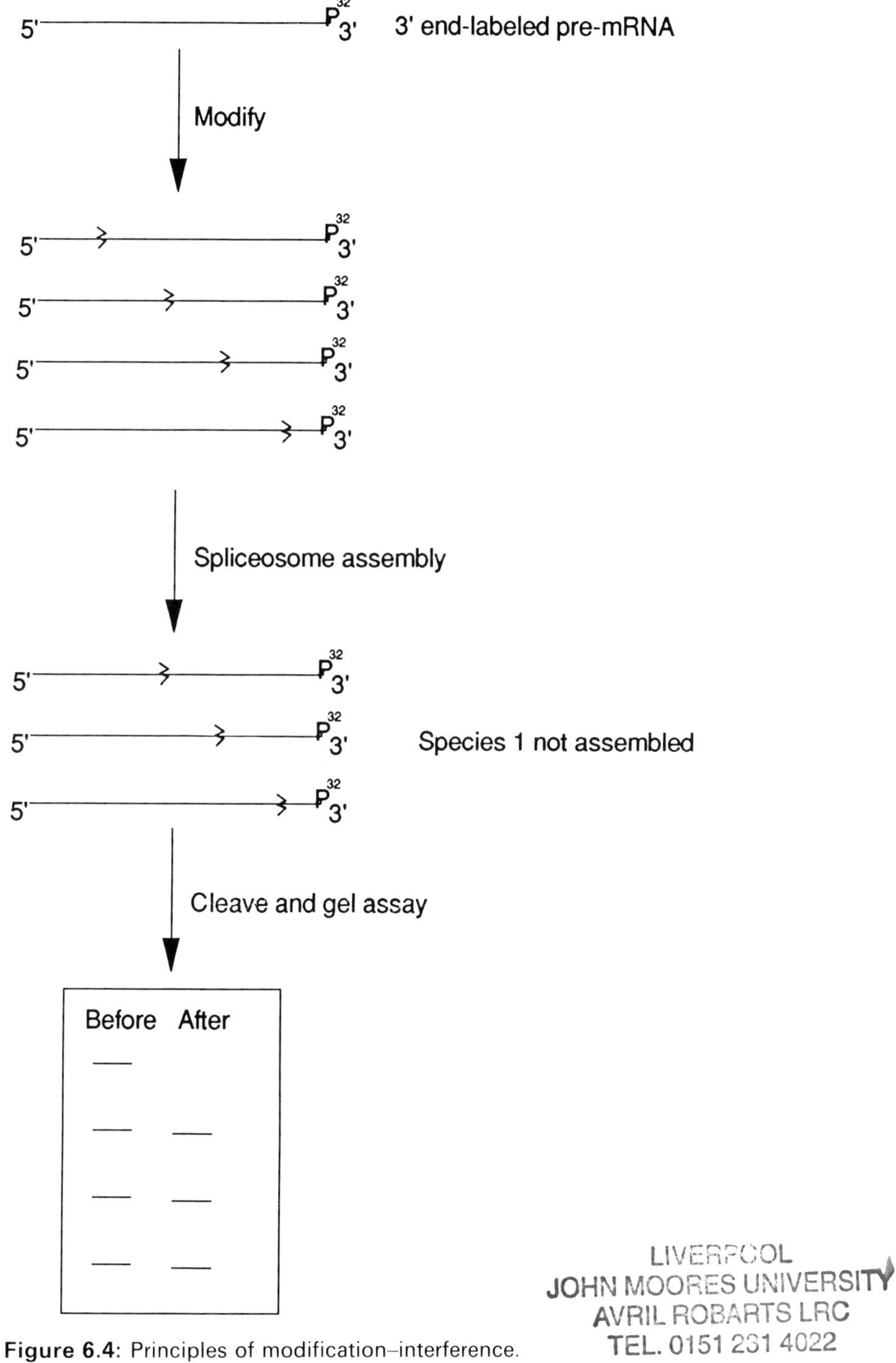

Figure 6.4: Principles of modification–interference.

LIVERPOOL
JOHN MOORES UNIVERSITY
AVRIL ROBARTS LRC
TEL. 0151 231 4022

proceeds for 30 min on ice. The reaction is terminated by addition of 200 µl 0.3 M sodium acetate (pH 3.8) and 750 µl ethanol. The RNA is collected by centrifugation at 4°C for 20 min. The RNA is redissolved in 200 µl 0.3 M sodium acetate (pH 3.8) and 600 µl ethanol is added. The precipitated RNA is collected by centrifugation as before.

(ii) For modifying U residues the RNA is redissolved in 10 µl water and 10 µl anhydrous hydrazine is added. Modification proceeds for 10 min on ice. The reaction is terminated by addition of 200 µl 0.3 M sodium acetate (pH 3.8) and 750 µl ethanol. The RNA is collected by centrifugation at 4°C for 20 min. The RNA is redissolved in 200 µl 0.3 M sodium acetate (pH 3.8) and 600 µl ethanol is added. The precipitated RNA is collected by centrifugation as before.

(iii) For modifying both C and U residues the RNA is redissolved in 20 µl fresh anhydrous hydrazine, 0.5 M NaCl. Modification proceeds for 30 min on ice. The reaction is terminated by addition of 200 µl 0.3 M sodium acetate (pH 3.8) and 750 µl ethanol. The RNA is collected by centrifugation at 4°C for 20 min. The RNA is redissolved in 200 µl 0.3 M sodium acetate (pH 3.8) and 600 µl ethanol is added. The precipitated RNA is collected by centrifugation as before.

3. Binding of protein to RNA. This carried out as described in Section 6.3.2 using the RNA and protein of interest.
4. Aniline cleavage of the RNA. The cleaved fragments are separated from the unreacted RNA molecules by gel electrophoresis, recovering the cleaved RNA molecules by elution. These RNAs are ethanol precipitated and collected by centrifugation. The pellet is washed with 70% ethanol and dried. The RNA is redissolved in 20 µl 1 M aniline and cleavage occurs at 60°C for 20 min in the dark. The reaction is terminated by RNA precipitation (by addition of 1.4 ml 1-butanol and vortexing). The RNA is collected by centrifugation at room temperature for 10 min. The RNA is then redissolved in 150 µl 1% (w/v) SDS. A second RNA precipitation (by addition of 1.4 ml 1-butanol and vortexing) is carried out and the RNA is collected by centrifugation at room temperature for 10 min. The pellet is washed with 1 ml 100% ethanol and dried.
5. Gel electrophoresis. The RNA fragments can be separated on a polyacrylamide sequencing gel (40 cm long 6% (w/v) polyacrylamide) at 30 V/cm for 1.5 h. The gel can then be exposed to X-ray film, generating a pattern similar to Figure 6.4.

Binding-site mutagenesis. Binding-site mutagenesis can be carried out with any type of RNA–protein interaction. RNA point mutants are generated by *in vitro* transcription from DNA templates or can be directly produced with oligonucleotide synthesizers (when small RNA molecules are being studied, e.g. snRNA). The templates vary at the sequence corresponding to the binding site of the RNA molecule to be produced. Large point mutant templates can be synthesized by PCR (Section 5.5 [11]) incorporating a transcription site into the 5′ terminus of the PCR primers, allowing RNA synthesis from the resultant PCR products.

Binding procedures as described in Section 6.3.2 are carried out with equal amounts of protein and RNA where the RNA in question is either unaltered or mutagenized. This type of study has been used to study all types of RNA–protein interaction. Assessment of binding is carried out by comparing the binding between the unaltered and the mutant reactions. This can be achieved by gel retardation (Section 6.3.3), filter/membrane binding assays (Section 6.3.3), or ribonuclease T1 digestion of the complex to depict proteins bound to RNA. Ribonuclease T1 analysis of binding sites [12] involves UV cross-linking of the complex, ribonuclease digestion and resolution of proteins bound to the protected radiolabeled RNA fragments in polyacrylamide gels.

6.3.5 Definition of RNA species essential to protein–RNA complex function

To ascertain which members of a complex: RNA–protein assembly are essential to the normal function, selective removal of a certain component of the complex is performed. This type of experiment is normally used in spliceosome investigations where cell extracts are being used. Simple interactions can be studied by synthesis of the elements that are desired in the reaction without the need for cell extracts. Complex interactions, however, use cell extracts as not all the functional members have yet been defined.

There are a few selective removal methods applied to these studies which include: RNase H digestion analysis and protection–immunoprecipitation.

Ribonuclease H digestion. RNase H will digest RNA molecules if they are bound to DNA oligonucleotides. Thus, specific hybridization of a designed oligonucleotide to the RNA that is to be removed from a mixture of RNAs (and therefore the subsequent binding reaction) can mediate removal of the RNA by specific digestion. Detailed protocols for this procedure have been described [13].

Protection–immunoprecipitation. Protection–immunoprecipitation analysis is usually a two-step procedure where a protein–RNA complex is subject to a RNase digestion which degrades any RNA not involved in the interaction (protection). The second step involves the use of an antibody against one of the proteins in a multiprotein complex. For example, in a model system, the transcript of 28S rRNA can be radiolabeled during *in vitro* synthesis and incubated with nuclear extract. The complex is digested with RNase T1 before the immunoprecipitation step. Antibodies isolated from autoimmune mice against (U3)RNP can be used to bind to their cognate partners. The antibodies are preattached to Sepharose beads which allows collection by centrifugation. This precipitation 'pulls down' the antibody and the protein of interest from the nuclear mixture. If the protein is essential to the binding reaction it will not appear as a 28S protein complex using polyacrylamide gel electrophoresis.

6.4 Protein synthesis

The ultimate test for mRNA is whether it will direct protein synthesis correctly. Cell-free systems are also used to investigate other aspects of RNA function such as splicing (snRNA and rRNA), translation (tRNA–tRNA synthetase interactions). Cell-free systems for bacterial and eukaryotic protein synthesis have been available for many years; only organelle protein synthesis has proved difficult to dissect into its components. Cell extracts provide the complete apparatus required for translation which consists of so many individual elements as to make it nearly impossible to create by individual *de novo* synthetic procedures.

Of all the cell-free systems, the rabbit reticulocyte lysate and the wheat germ extract systems have been most popular. Both have a number of advantages over other systems and are easy to prepare, provided that good laboratory practice is adopted to avoid contamination with ribonucleases. These cell free systems are commercially available, avoiding the need for preparation. One limitation of the reticulocyte lysate and wheat germ systems is the lack of endoplasmic reticulum for correct processing of translation products that contain signal peptides. This can be overcome by the addition of canine pancreatic microsomal membranes to a standard translation reaction [14].

Post-translational modifications of proteins are of interest in themselves. A shift in size after modification can be visualized by gel electrophoresis.

6.4.1 Rabbit reticulocyte translation system

The most popular system by far is that derived from rabbit reticulocytes using hypotonic lysis followed by high-speed centrifugation to remove the cell membranes. The use of the rabbit reticulocyte system is often favored for translation of larger mRNA species, and is generally recommended when microsomal membranes are to be added to facilitate post-translational processing of translation products.

The rabbit reticulocyte lysate system needs to be treated to allow more efficient utilization of exogenous mRNA. Therefore, the endogenous globin mRNA is degraded by incubation with Ca^{2+}-dependent micrococcal nuclease that can subsequently be inactivated by chelation with EGTA [15]. Such nuclease-treated lysates are commercially available.

Protocol 6.2: *In vitro* translation using the rabbit reticulocyte system

Protein synthesis is carried out with *in vitro* synthesized RNA and reticulocyte lysate. The lysate is supplemented with amino acids and any metal cofactors necessary for correct protein folding. Two reactions are performed side by side; the first, a radiolabeling reaction, allows quantitation of the synthesis of protein, whilst the other reaction is preparative.

1. 400 µl lysate is thawed and supplemented with 8 µl 1 mM amino acid mixture (all amino acids except methionine) and 4 µl 10 mM $ZnCl_2$ (if protein folding needs zinc ions, e.g. certain DNA binding proteins)[16].
2. Protein synthesis is initiated by mixing 20 µl (1–2 µg) of *in vitro* synthesized RNA and 40 µl supplemented lysate.
3. 18 µl of this translation mix is transfered to a sterile 0.5 ml microcentrifuge tube before adding 2 µl of L-[^{35}S]methionine (37 TBq (1000 Ci)/mmol at 555 MBq (15 mCi/ml)).
4. The remaining 22 µl translation mix is supplemented with 2 µl unlabeled 1 mM methionine. These two reactions are incubated at 30°C for up to 1 h.
5. Synthesis is assessed by spotting 20 µl aliquots of the radiolabeled mix on to 25 mm diameter filter paper or glass fiber disks.
6. To precipitate the protein, dried disks are immersed in ice-cold 10% (w/v) trichloroacetic acid (TCA) for 10 min. The disks are then immersed in hot 10% TCA for 10 min. Two ethanol washes, a wash with ethanol:ethyl ether (1:1 v/v) and a final wash with ethyl ether removes non-precipitated material.
7. The amount of newly synthesized, precipitated protein is assessed by scintillation counting of the dried disks.

6.4.2 Wheat germ extract translation system

The other cell-free system still in general use is that derived from the postmicrosomal supernatant of wheat germ [14], which has an intrinsically low content of endogenous mRNA. Wheat germ extract readily translates certain RNA preparations, such as those containing low concentrations of double-stranded RNA or oxidized thiols, which are inhibitory to reticulocyte lysate.

Cell-free extracts of wheat embryos catalyze the incorporation of radioactive amino acids into protein in response to several mRNAs.

Protocol 6.3: *In vitro* translation using the wheatgerm system

1. Wheat germ extract is prepared by grinding the wheat germ in a cooled mortar with acid-washed, autoclaved sand and the extract is centrifuged at 30 000 *g* for 10 min at 5°C.
2. The supernatant is gel filtered over Sephadex G-25 and then stored frozen in liquid nitrogen.
3. The extract is mixed with a master mix which has a similar composition to that of the rabbit reticulocyte system. Reaction mixtures are then incubated for 30 min at 30°C.
4. The reaction is stopped by adding 0.2 ml 16% (w/v) trichloroacetic acid (TCA) containing unlabeled leucine followed by 4 ml 5% TCA. The suspension is cooled on ice for 10 min and the insoluble material is collected by centrifugation.
5. The pellet is then suspended in 4.0 ml 5% TCA containing 6.5 μmol leucine, and heated for 15 min at 90°C. After 8 min on ice, the insoluble material is collected on glass fiber disks. The disks are washed twice with 5% TCA, dried under an infrared lamp, and the radioactivity of each filter is measured in a liquid scintillation counter.

One limitation of the reticulocyte lysate and wheat germ systems is the lack of endoplasmic reticulum for correct processing of translation products that contain signal peptides. This may be overcome by the addition of canine pancreatic microsomal membranes to a standard translation reaction [14]. Processing events are generally detected as shifts in the apparent molecular weight of translation products and can readily be detected by electrophoresis on SDS polyacrylamide gels.

6.5 RNA–RNA complexes

The discovery of RNA catalysis has led to intensive study into RNA–RNA complex formation and function. RNA catalysis by 'ribozymes' encompasses specific nuclease reactions, self-splicing introns and some polymerizing activities. The wide diversity of techniques used in this area of research precludes a detailed description of protocols used for all the available techniques and instead an overview of the available techniques is given, together with references to the methodologies.

6.5.1 Nuclease action

RNase P is an enzyme containing protein and RNA where the RNA (M1 RNA), when synthesized *in vitro* can catalyze the cleavage of precursor $tRNA^{Tyr}$ (pTyr) into $tRNA^{Tyr}$. The protein part (C5) of the enzyme is thought to twist the RNA into the conformational state that is enzymatically active *in vivo* [17].

RNA catalysis studies usually define the catalytic action by comparing the substrate before and after catalysis. For example, cleavage reactions can be studied by gel electrophoresis.

Investigation of cleavage reactions using gel electrophoresis. Cleavage reactions produce a shortened RNA molecule which can be visualized by electrophoretic procedures. Reactions can be stopped by addition of denaturing gel loading buffer of reaction aliquots over time. These samples can be electrophoresed through agarose (3%) or polyacrylamide (10–20%) gels. Visualization with EtBr allows comparisons of the reaction products over time.

Investigation of active sites by mutational studies. RNA molecules may be synthesized *in vitro* using DNA templates and specific RNA polymerases (Section 3.5.1). The DNA templates can be altered at positions that correspond to areas of interest in the RNA molecule to be produced. Specific base changes can be accomplished by the polymerase chain reaction (Section 5.5). The RNA enzymatic activities can then be assessed.

6.5.2 Self-splicing activity

There are two major types of self-splicing introns (group I and II) which can catalyze their own excision under certain circumstances [18] (see *Figure 1.5*). These RNA molecules are thought to be derived from mobile genetic elements that had an excision and protein coding function. The functionality of these RNA elements has in many cases degraded such that efficient splicing is now only possible with protein cofactors.

Group I introns. These introns, first found in a *Tetrahymena* rRNA, splice in a two-step reaction. The first esterification is initiated by a nucleophilic attack by a guanosine upon the 5′ splice site. The released 5′ exon then attacks the 3′ splice site resulting in intron release.

These reactions have been studied with a variety of techniques including cleavage reactions, deoxynucleotide/phosphorothioate substitution, mutant analysis, specific cleavage analyses and competitive inhibition of the guanosine binding site.

Group II introns. These have only been found in organelles such as fungal and plant mitochondria and chloroplasts. There is a remarkable similarity between the splicing mechanism of group II introns and nuclear mRNA introns, suggesting that the latter evolved from the former perhaps as a result of the introns moving as genetic elements from a symbiotic bacteria (organelle) to the host genome. Later evolution of these sequences in eukaryotic cells is thought to have led to losses of function, necessitating protein cofactors to form spliceosomes around the RNA molecules (snRNAs).

Splicing in group II introns involves the formation of an intron lariat where the 5′ end of the intron forms a linkage to a nucleoside (usually an adenosine residue) by a 2′–5′ phosphodiester bond. The 5′ exon then attacks the 3′ end of the intron joining to the 3′ exon splicing the intron out.

Investigation of the function of group II introns has also used the methods just listed which are summarized next.

Non-specific cleavage. The chemical agent Fe(II)EDTA cleaves the sugar–phosphate backbone of the RNA molecule [19,20]. This has allowed structural features to be elucidated. It has revealed that magnesium ions are necessary for correct functional folding of the intron, that RNA domains fold in a well-defined order as the magnesium concentrations was raised and that a tertiary structure normally forms a buried catalytic core not in contact with surrounding solvent. It is thought that helices form in the RNA molecules and that these helices often lie upon one another in ‘stacks’.

Deoxynucleotide/phosphorothioate substitution. Incorporation of deoxynucleotides or phosphorothioate residues into RNA molecules has provided interesting structural information. Substituting does not have any effect unless residues are important for splicing [21]. Substitution of three of the groups in the *Tetrahymena* intron (U305, A306 and A308) resulted in a decrease in splicing when modified.

Mutational analysis. Binding-site mutagenesis can be carried out with any type of RNA–RNA interaction. RNA point mutants can be generated by *in vitro* transcription from DNA templates or can be directly produced with oligonucleotide synthesizers (when small RNA molecules are being studied, e.g. synthetic introns). The templates vary at the sequence corresponding to the binding site or functional RNA loops of the RNA molecule to be produced. Large point mutant templates can be synthesized by PCR (Section 5.4 and [11]) incorporating a transcription site into the 5′ terminus of the PCR primers. For example, mutation of a group I intron of the thymidylate synthase (*td*) gene of the bacteriophage T4 at the conserved P7 secondary structure led to an alteration in affinity of the intron for the guanosine residue necessary for splicing.

Specific cleavage analyses. RNase H will digest RNA molecules if they are bound to DNA oligonucleotides [13, 22]. Thus, specific hybridization of a designed oligonucleotide to the RNA that is to be removed from the subsequent reaction can mediate removal of the RNA by specific digestion. Nuclear extracts are used which have endogenous levels of RNase H. These are incubated with an excess of oligonucleotide (10 ng/μl extract) to ensure that all target RNAs form hybrids in a solution containing 2.7 mM $MgCl_2$. Hybridization occurs at a temperature slightly below the melting temperature of the hybrid for 15 min. The ribonuclease is then inhibited by chelating the magnesium ions in the solution. This is achieved by addition of EDTA to a final concentration of 20 mM.

Competitive inhibition of the guanosine-binding site. The guanosine binding site of group I introns mediates the initiation of cleavage reactions by binding a guanosine residue which can discriminate between nucleosides. Guanidino groups possessed by arginine or antibiotics such as streptomycin can compete for the binding site acting as competitive inhibitors of splicing [23]. This inhibition can only be applied to group I introns.

6.5.3 Polymerase-like activities of RNA

The polymerizing abilities of RNA molecules is seen in certain circumstances by group I introns. *Figure 6.5* depicts the possible reactions catalyzed by ribozymes. These reactions use the splicing site to accomplish different end-products under different conditions. The *Tetrahymena* ribozyme has been shown to catalyze the reverse of an aminoacylation reaction where a modified RNA internal guide sequence allowed the hydrolysis of an aminoacyl-ester, *N*-formyl-L-methionine that was attached to the acceptor stem (CCA sequence). Thus, ribozymes could have carried out primitive versions of aminoacylation in early systems.

Polymerase activities have been seen in non-templated addition of nucleotides to oligonucleotides bound to the internal guide sequence – a nucleotidyl transferase function/ligation. Template-dependent action (the addition of oligonucleotides), necessary for evolutionary survival, was demonstrated in a reaction that can be thought of as a reversal of splicing [24]. The ability to self-replicate has also been seen when different segments of a group I ribozyme formed a subunit ribozyme that replicated one of its own segments by a template-dependent and directed ligation of oligonucleotides [25].

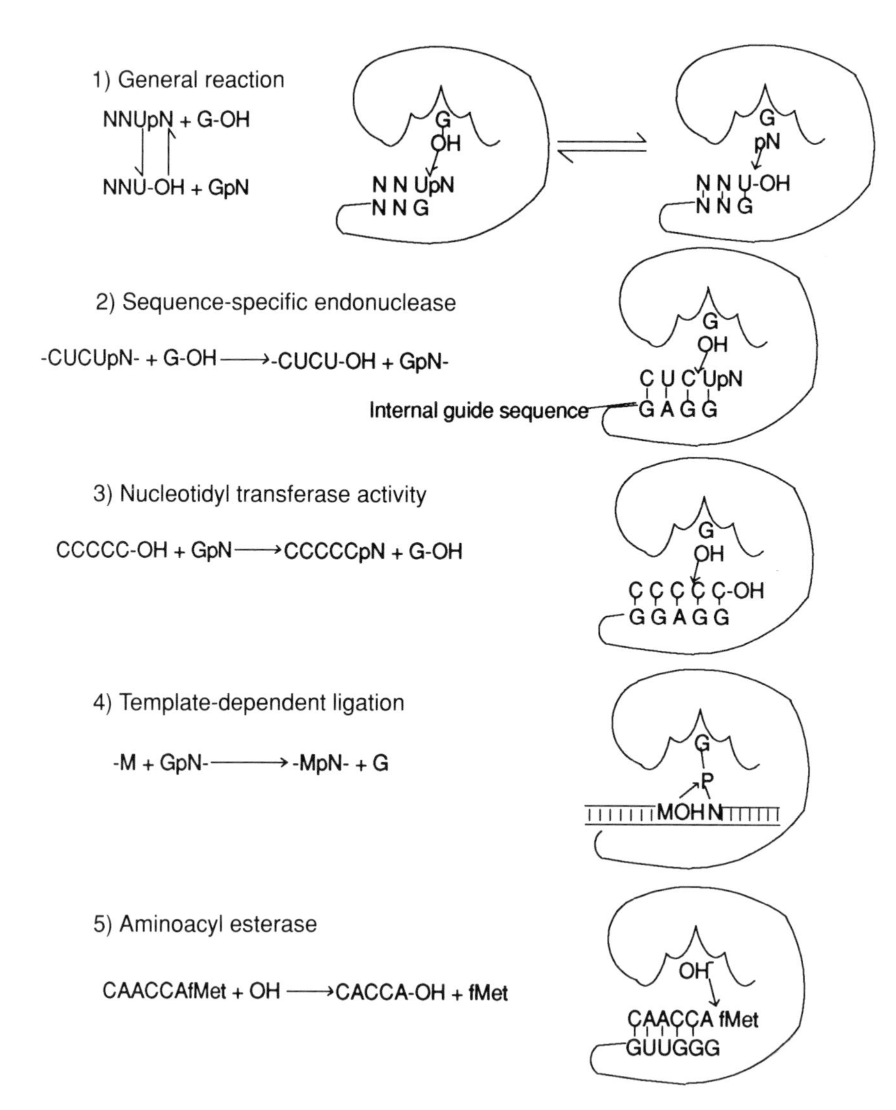

Figure 6.5: Possible reactions catalyzed by ribozymes.

References

1. **Michel, F. and Westhof, E.** (1990) *J. Mol. Biol.*, **216**, 585.
2. **Zapp, M.L., Stern, S. and Green, M.R.** (1993) *Cell*, **74**, 969.
3. **Jaffery, S.R., Haile, D.J., Klausner, R.D. and Harford, J.B.** (1993) *Nucl. Acids Res.*, **21**, 4627.
4. **Siomi, H., Siomi, M.C., Nussbaum, R.L. and Dreyfuss, G.** (1993) *Cell*, **74**, 291.
5. **Mundus, D.A., Bulygin, K.N., Yamkovoy, V.I., Malygin, A.A., Repkova, M.N., Vratskikh, L.V., Venijaminova, A.G., Vladimirov, S.N. and Karpova, G.G.** (1993) *Biochem. Biophys. Acta*, **1173**, 273.
6. **Jackson, V.** (1978) *Cell*, **15**, 945.
7. **Ip, Y.T., Jackson, V., Meier, J. and Chalkley, R.** (1988) *J. Biol. Chem.*, **263**, 14044.

8. **Draper, D.E., Deckman, I.C. and Vartikar, J.V.** (1988) *Meth. Enzymol.*, **164**, 203.
9. **Ruby, S.W., Geolz, S.E., Hostomsky, Z. and Abelson, J.N.** (1990) *Meth. Enzymol.*, **181**, 97.
10. **Conway, L. and Wickens, M.** (1989) *Meth. Enzymol.*, **180**, 369.
11. **Clackson, T., Güssow, D. and Jones, P.T.** (1991) in *PCR: a Practical Approach* (M.J. Mcpherson, P. Quirke and G.R. Taylor, eds). IRL Press, Oxford. pp. 187.
12. **Aebi, M.** (1990) *Meth. Enzymol.*, **181,** 43.
13. **Krämer, A.** (1990) *Meth. Enzymol.*, **181,** 284.
14. **Clemens, M.J.** (1987) in *Transcription and Translation: a Practical Approach* (B.D. Hames and S.J. Higgins, eds). IRL Press, Oxford. pp. 231.
15. **Sambrook, J., Fritsch, E.F. and Maniatis, T. (eds)** (1989) *Molecular Cloning – A Laboratory Manual,* 2nd edn. Cold Spring Harbor Laboratory Press, New York. Section 18.76.
16. **White, R. and Parker, M.** (1993) in *Transcription Factors – a Practical Approach* (D.S. Latchman, ed.) IRL Press, Oxford.
17. **Altman, S., Baer, M., Gold, H., Guerrier-Takada, C., Kirsebom, L., Lawrence, N., Lumelsky, N. and Vioque, A.** (1987) in *Molecular Biology of RNA: New Perspectives* (M. Inouye and B.S. Dudock, eds). Academic Press, London.
18. **Saldanha, R., Mohr, G., Belfort, M. and Lambowitz, A.M.** (1993) *FASEB J.*, **7**, 15.
19. **Cech, T.R.** (1990) *Annu. Rev. Biochem.*, **59**, 543.
20. **Celander, D.W. and Cech, T.R.** (1991) *Science,* **251**, 401.
21. **Waring, R.B.** (1989) *Nucl. Acids Res.*, **17**, 10281.
22. **Konarska, M.M. and Sharp, P.A.** (1987) *Cell,* **49**, 763.
23. **von Ahsen, U. and Schroeder, R.** (1991) *Nucl. Acids Res.*, **19**, 2261.
24. **Doudna, J.A. and Szostak, J.W.** (1989) *Nature,* **339**, 519.
25. **Doudna, J.A., Couture, S. and Szostak, J.W.** (1991) Science, **251**, 1605.

7 Isolation and analysis of ribonucleoproteins

7.1 Ribonucleoproteins (RNPs)

In the cell, virtually all types of RNA are associated with proteins to a greater or lesser extent. Some RNPs, such as ribosomes, have a precise composition in terms of their RNA and protein components while other particles such as mRNP are much more heterogeneous. Section 1.5 summarizes the different types of nucleoproteins found in prokaryotic and eukaryotic cells. A number of practical manuals covering all aspects have been published on RNPs [1–3]. This chapter is restricted to describing some of the important methods used for the isolation and fractionation of RNPs together with some simple analytical methods that can be used for their characterization.

7.2 Isolation of ribonucleoproteins

7.2.1 Homogenization methods

All methods used for isolating RNPs can be divided into two stages; namely extraction and subsequent purification. For example, ribosomes may be released from cells simply by breaking open the cell, while in the case of other complexes more sophisticated procedures are necessary to release the complexes from cellular structures. Homogenization can be achieved as described in Section 2.2 by osmotic lysis, mechanical shearing or enzymatic digestion depending on the type of cells being used.

An important factor during homogenization is the composition of the medium. The content of ions, especially Mg^{2+}, in the homogenization medium is of paramount importance since these ions are crucial in maintaining the composition and function of many RNPs. As a rule, the concentration of monovalent cations should be about 10 times that of Mg^{2+}, where the Mg^{2+} concentration is usually 5 mM. In addition, the overall ionic strength of the medium is important as it determines

Table 7.1: Homogenization buffers used in ribosomal isolation procedures.

Buffer composition	Source of ribosomes
20 mM Tris-HCl (pH 7.5 at 4°C), 10.5 mM magnesium acetate, 100 mM NH_4Cl, 0.5 mM EDTA, 3 mM 2-mercaptoethanol	Prokaryotes
50 mM Tris-HCl (pH 7.6 at 20°C), 25 mM KCl, 5 mM $MgCl_2$, 0.25 M sucrose	Mammalian liver
5 mM Tris-HCl (pH 7.5 at 20°C), 1 mM EDTA, 340 mM sucrose	Mammalian mitochondria
10 mM Tris-HCl (pH 7.6 at 26°C), 50 mM KCl, 10 mM magnesium acetate, 0.7 M sorbitol, 7 mM 2-mercaptoethanol	Spinach chloroplasts

the protein components found in the complex. This ionic concentration also defines the number of 'tight couples' of ribosomes which are resistant to subsequent centrifugal fractionation and are able to synthesize protein. *Table 7.1* details some homogenization buffers used in ribosome isolation.

Once released the subsequent steps for purification of the nucleoproteins depend on the nature of the complexes and the contaminants present.

7.2.2 Preparation of ribosomes, ribosomal subunits and polysomes

All cells contain large numbers of ribosomes, and homogenization of cells releases a large proportion of the ribosomes into the homogenization medium [4]. As with the isolation of RNA, it is important to avoid degradation of ribosomes during isolation procedures. Rapid handling at low temperature is usually sufficient to maintain ribosomal integrity. Cellular debris can be pelleted by centrifugation of the homogenate at 15 000 *g* for 10 min, leaving free ribosomes in the supernatant. Ribosomes (and organelles) can then be pelleted by centrifugation of the recovered supernatant at 100 000 *g* for 60 min. All preparatory steps should be carried out at 0–5°C to minimize degradation.

Protocol 7.1: Preparation of prokaryotic ribosomes

1. Cells may be homogenized in a French press or by grinding in alumina in homogenization buffer containing 20 mM Tris-HCl (pH 7.4 at 4°C), 10.5 mM magnesium acetate, 100 mM ammonium chloride, 0.5 mM EDTA and 3 mM 2-mercaptoethanol.
2. The majority of the cellular debris can be pelleted by centrifugation at 30 000 *g* for 45 min at 0–5°C. The resultant supernatant is often referred to as the 'S30' supernatant.
3. The preparation of ribosomes will be purer if they are sedimented through a cushion of 1.1 M sucrose in buffer containing 20 mM Tris-HCl (pH 7.4 at 4°C), 10.5 mM magnesium acetate, 0.5 M ammonium chloride, 0.5 mM EDTA, 3 mM 2-mercaptoethanol.

The sucrose solution is pipetted into the bottom of the centrifuge tube and the S30 supernatant (step 2) is layered over it. The density of this cushion allows the ribosomes to pass through but stops other cellular components of lower density. As long as the preparation is carried out carefully, to avoid degradation by ribonucleases, a large proportion of the pelleted ribosomes will be in the form of polysomes, that are still associated with mRNA. Sucrose/salt-washed ribosomes are pelleted by centrifugation at 100 000 *g* for 15 h at 15°C.

4. The pellet is resuspended by gentle brushing of the pellet with a glass rod and the sucrose/salt washing is repeated (step 3).
5. The pellet is resuspended in buffer containing 10 mM Tris-HCl (pH 7.5 at 4°C), 5.25 mM magnesium acetate, 60 mM ammonium chloride, 0.25 mM EDTA and 3 mM 2-mercapto-ethanol. The resuspended pellet is dialyzed against this buffer for 6 h at 4°C, changing the dialysis buffer every 2 h.
6. Ribosomes can be removed from the mRNA by incubating them at 37°C for 30 min, allowing the ribosomes to 'run off' the mRNA.
7. Treatment of the ribosomes, by dialyzing in buffer containing 10 mM Tris-HCl (pH 7.5), 1.1 mM magnesium acetate, 60 mM ammonium chloride, 0.1 mM EDTA and 2 mM 2-mercaptoethanol for 6 h at 4°C, will cause the two subunits to dissociate.
8. The subunits can then be separated by rate-zonal centrifugation on a 15–30% (w/v) sucrose gradient. To do this 50–60 A_{260} units of separated subunits are layered on to 34 ml of 15–30% (w/v) sucrose gradients made up in dialysis buffer (step 7) by centrifugation at 43 000 *g* for 16 h at 4°C (*Figure 7.1*). Note that a rough calculation of ribosome concentration can be achieved by equating 10 A_{260} units to 1 mg/ml ribosomes.
9. The gradient is fractionated and the 30S and 50S peak fractions can be located by UV absorption and pooled. The magnesium concentration is raised to 10 mM and the subunits are ethanol precipitated (addition of 0.65 vol. ice-cold ethanol, −20°C for 2 h and pelleting of precipitation by centrifugation at 16 000 *g* for 15 min at 4°C).

Protocol 7.2: Preparation of eukaryotic ribosomes

1. Homogenize tissue in 4 ml per gram of tissue in ice-cold homogenizing buffer containing 0.25 M sucrose, 50 mM Tris-HCl (pH 7.6 at 20°C), 25 mM KCl, 5 mM $MgCl_2$ by cutting finely and homogenizing with three strokes in a loosely fitted Teflon-glass homogenizer and five or six strokes with a motor-driven Teflon pestle at 1000 r.p.m.
2. Mitochondria can be removed by pelleting at 15 000 *g* for 10 min at 4°C.

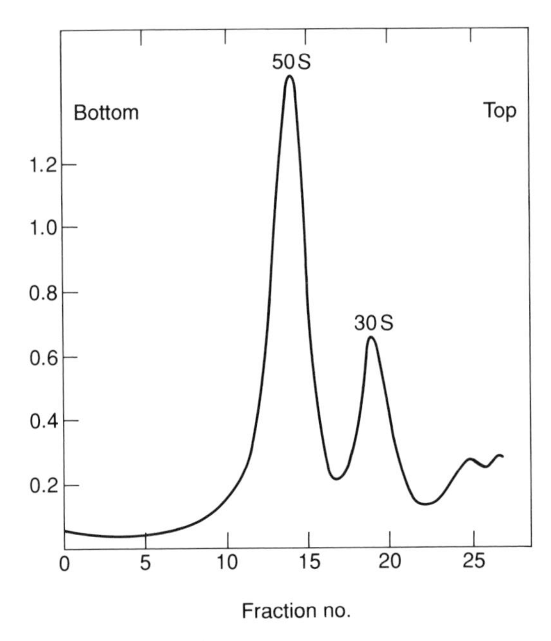

Figure 7.1: Separation of prokaryotic 50S and 30S ribosomal subunits by rate-zonal centrifugation on a 15–30% sucrose gradient.

3. The top two-thirds of the supernatant is recovered and stored on ice. The pellet is then re-extracted by the addition of one volume of homogenization buffer followed by homogenization with three strokes with a motor-driven Teflon pestle at 1000 r.p.m. Centrifugation is carried out as in step 2 and the second supernatant is combined with the stored supernatant.
4. The combined supernatants are filtered through four layers of muslin before the filtrate is centrifuged at 145 000 *g* for 2 h to produce a microsomal pellet.
5. The pellet is resuspended in an equal volume of buffer containing 0.15 M sucrose, 35 mM Tris-HCl (pH 7.8), 25 mM KCl, 10 mM $MgCl_2$, 0.15 M sucrose, 1 mM DTT by gentle stirring with a glass rod. The pellet often tends to be rather sticky and so this step can take some time.
6. The concentration of potassium chloride is raised to 0.5 M by adding 2.5 M KCl, 10 mM $MgCl_2$. Next, a tenth volume of 10% sodium deoxycholate in 0.5 M KCl is added to dissolve the membranes associated with the ribosomal material. Mix the suspension well after addition of sodium deoxycholate.
7. The suspension is layered over an equal volume of 1.0 M sucrose (cushion) in homogenization buffer. The ribosomal fraction will pellet through the cushion by centrifugation at 176 000 *g* for 90 min. The pellet is rinsed in buffer containing 0.25 M sucrose, 50 mM Tris-HCl (pH 7.8), 50 mM KCl, 5 mM

$MgCl_2$, 6 mM 2-mercaptoethanol, 10 mM potassium bicarbonate and resuspended by gently brushing with a glass rod into a volume of the same buffer to give a concentration of about 20–30 mg rRNA/ml.

8. Aggregates and debris can be removed from the resuspended pellet by centrifugation at 20 000 *g* for 10 min. At this stage most of the ribosomes are present as polysomes, that is complexes of ribosomes and mRNP, and these can be separated by rate-zonal centrifugation on a 15–30% sucrose gradient at 100 000 g for 2.5 h at 5°C (*Figure 7.2*).
9. The concentration of KCl is raised to 0.5 M and freshly prepared puromycin is added to a final concentration of 0.5 mM. Incubation at 37°C for 15 min ensures mRNA 'run off' has occured.
10. Addition of suitable volumes of 2.5 M KCl, 1 M $MgCl_2$ and 0.1 M 2-mercaptoethanol to final concentrations of 1 M, 10 mM and 20 mM, respectively, causes the ribosomes to dissociate.
11. The subunits can then be separated by rate-zonal centrifugation on a 15–30% (w/v) sucrose gradient in buffer containing 50 mM Tris-HCl (pH 7.6), 850 mM KCl, 10 mM $MgCl_2$, 10 mM 2-mercaptoethanol. The dissociated subunits are layered on to 15–30% (w/v) linear sucrose gradients made up in the same buffer followed by centrifugation at 95 000 *g* for 5 h at 5°C.
12. The gradient is fractionated and the 40S and 60S subunit peak fractions can be located and pooled.
13. The subunits are collected by centrifugation at 130 000 *g* for 12 h at 4°C and the pellet is resuspended in buffer containing 20 mM Tris-HCl (pH 7.6), 100 mM KCl, 5 mM $MgCl_2$, 10 mM 2-

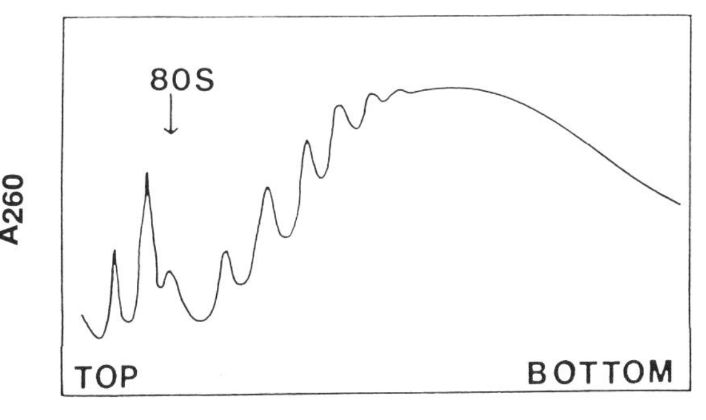

Figure 7.2: Separation of rat liver polysomes on a 15–30% sucrose gradient. The sedimentation position of the 80S monomeric ribosomes is shown, the two peaks sedimenting more slowly represent the ribosomal subunits and those sedimenting faster are the polysomes. Polysomes containing up to six ribosomes are clearly identifiable.

mercaptoethanol and 0.25 M sucrose at a concentration of 100–150 A_{260} units/ml.

It has been shown that it is possible to replace some of the centrifugation steps in isolation procedures by use of gel filtration [5]. This method yields active ribosomes that are virtually free of soluble protein synthesis factors such as elongation factor G.

Protocol 7.3: Fractionation of crude homogenates using gel filtration

1. Gel filtration purification is best carried out by using Sephacryl column chromatography [5]. Bacterial cells can be washed in buffer containing 10 mM HEPES-KOH (pH 7.5), 10 mM $MgCl_2$, 70 mM KCl and 3 mM 2-mercaptoethanol and collected by centrifugation (4000 g for 15 min).
2. The bacterial pellet (50 g) can then be resuspended in 100 ml homogenizing buffer containing 20 mM HEPES-KOH (pH 7.5), 5 mM $MgCl_2$, 0.5 M $CaCl_2$, 8 mM putrescine, 5 mM NH_4Cl, 95 mM KCl, 1 mM dithioerythritol (DTE), 1 mM spermidine and 3 mM 2-mercaptoethanol. Cells may then be lysed with two passes through a French press. Cell debris can then be pelleted by centrifugation at 28 000 *g* for 30 min. The supernatant should be recovered and residual debris pelleted by repeating the centrifugation at 28 000 *g* for 30 min.
3. Finely powdered solid ammonium sulfate (210 mg/ml) is gradually added to the recovered supernatant whilst stirring. Ammonium hydroxide is then used to adjust the pH to 7.5 and protein precipitation occurs over 30 min at 5°C. Centrifugation at 28 000 *g* for 30 min pellets the protein fraction, which is discarded.
4. The supernatant containing the ribosomes may then be used for the gel filtration. A Sephacryl S300 column (9 × 135 cm, Pharmacia) is pre-equilibrated in buffer containing 20 mM HEPES-KOH (pH 7.5), 5 mM $MgCl_2$, 0.5 M $CaCl_2$, 8 mM putrescine, 0.5 M NH_4Cl, 95 mM KCl, 1 mM DTE, 1 mM spermidine and 3 mM 2-mercaptoethanol.
5. The clear supernatant is loaded on to the Sephacryl S300 column and eluted at a flow rate of 30–100 ml/h.
6. When the fractions are analyzed by zonal centrifugation, a typical preparation produces a well-defined peak (*Figure 7.3a*). Analysis of the peak eluting material shows that the peak is rich in 70S ribosomes (*Figure 7.3b*).
7. In general preparative methods, the first two-thirds of the large initial A_{260} peak are pooled and used as a purified polysome preparation without assaying the individual fractions for polysome content.

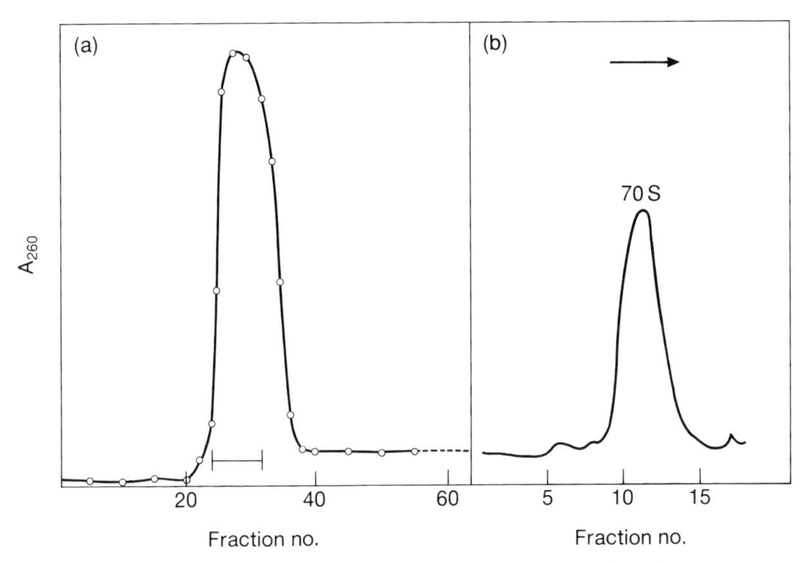

Figure 7.3: (a) Separation of bacterial ribosomes by gel filtration on Sephacryl S-200. Fractions from the peak region indicated were then pooled and analyzed by rate-zonal centrifugation on a 10–30% sucrose gradient (b). Derived from Ref. 5.

7.2.3 Preparation of mRNP complexes

The mRNA of eukaryotic cells can generally be found free in the cytoplasm or, as mentioned in the previous Section, complexed with ribosomes. The latter location is the site of the highest concentration of mRNA in most cell types. In addition, precursor mRNA mainly in the form of spliceosomes can also be found in the nucleus. In bacteria, ribosomes bind to nascent mRNA before transcription is finished and the mRNA is rapidly degraded within a few minutes of being translated.

Protocol 7.4: Preparation of mRNP complexes

1. Treatment of isolated polysomes with 1 mM puromycin can dissociate the ribosomal subunits and release mRNA molecules in the form of mRNP complexes.
2. Once dissociated, the mRNPs can be separated from the ribosomes and ribosomal subunits by isopycnic centrifugation in CsCl, Nycodenz or Metrizamide gradients [6], since the mRNPs tend to be richer in proteins and band at lower densities.

The mRNPs may be fractionated in a 30–50% Nycodenz gradient containing magnesium ions in a fixed-angle rotor at

100 000 *g* for 40 h. Ribosomes may be fractionated in gradients of 48% Nycodenz containing magnesium ions in a fixed-angle rotor at 100 000 *g* for 40 h. In the presence of magnesium ions, ribosomes band at densities close to 1.3 g/ml while the mRNP bands much lighter in Nycodenz. Some loosely bound proteins dissociate from ribosomes during centrifugation.

Alternatively, mRNP can be separated on CsCl gradients. However, in this case, prior to centrifugation on CsCl gradients the ribosomes must be fixed, usually with 1–4% formaldehyde (see Section 6.3.2). CsCl gradients with an initial density of 1.6 g/ml are generated by mixing CsCl with the sample and centrifuging for 16–18 h at 100 000 *g*. The CsCl gradient forms during centrifugation and ribosomal subunits reach positions corresponding to their equilibrium buoyant densities. The mRNP generally bands lighter (1.4 g/ml) than the ribosomes (1.55 g/ml).

3. Once isolated the formaldehyde cross-linking of the nucleoproteins can be reversed by incubation at high temperatures [7,8], allowing further analysis.
4. Alternatively, if only the RNA component is to be analyzed then the protein of the fixed nucleoprotein complex can be digested away with RNase-free protease. This can be achieved by incubating at room temperature for 15 min with 0.1 mg/ml proteinase K.

7.2.4 Preparation of spliceosomes

Ribonucleoprotein complexes are composed of pre-mRNA and nuclear proteins, and are intermediates of nuclear processing of mRNA. With the development of cell-free systems for the *in vitro* splicing of mRNA precursors, it has become feasible to analyze the isolated individual parts of the splicing machinery. Extensive research into this area is uncovering interesting mechanistic details.

Protocol 7.5: Preparation of spliceosomes

1. Nuclear extracts active in pre-mRNA splicing are prepared by a modification of the procedure originally reported by Dignam *et al.* [9].
2. Cultured cells are collected by centrifugation and washed in PBS. The pelleted cells are resuspended in cold PBS and centrifuged for 10 min at 5000 *g* (5°C).
3. The resulting pellets are swollen by resuspending them in homogenization buffer containing 10 mM HEPES-KOH (pH 7.9), 1.5 mM $MgCl_2$, 10 mM KCl, 0.5 mM DTT and the cells are then pelleted by centrifugation for 10 min at 1000 *g*.

4. Pellets are transfered with a small volume of this buffer into an ice-cold, tight-fitting Dounce homogenizer and the cells are broken with 15 bidirectional strokes of the pestle.
5. The homogenate is centrifuged at 1000 *g* for 10 min to pellet the nuclei.
6. The nuclei are then resuspended in nuclei buffer containing 20 mM HEPES-KOH (pH 7.9), 25% glycerol, 0.42 M NaCl, 1.5 mM $MgCl_2$, 0.2 mM EDTA, 0.5 mM phenylmethyl sulfonyl fluoride (PMSF) and 0.5 mM DTT and transferred to a Dounce homogenizer. The high salt concentration in the nuclei homogenization buffer precipitates the chromatin.
7. The nuclei are disrupted by 10 bidirectional strokes of the tight pestle and the suspension is stirred on ice for 30 min.
8. Nuclear debris is removed by centrifugation at 26 000 *g* for 30 min at 5°C.
9. The supernatant is dialyzed for 5 h on ice against two changes of 1 liter of dialysis buffer containing 20 mM HEPES-KOH (pH 7.9), 20% glycerol, 0.1 M KCl, 0.2 mM EDTA, 0.5 mM DTT.
10. After dialysis, the extract is cleared of insoluble material by centrifugation at 26 000 *g* for 30 min at 5°C. The supernatant is stored in aliquots at −80°C.

The resulting splicing extract can be further fractionated by chromatographic method into six fractions that have to be combined to yield splicing intermediates and reaction products [10]. Two of the active fractions contain the snRNPs whereas the remaining four appear to consist of protein factors. The major snRNPs U1–U6 can be considerably purified. In addition, it is also possible to separate the minor snRNPs U7 and U11 from the major snRNP particles. This chromatographic procedure is illustrated schematically in *Figure 7.4.*

Another approach, described in Section 6.3.3, is to use affinity chromatography with biotinylated RNAs as an alternative method for purifying spliceosomes and the components involved in splicing using pre-mRNA synthesized *in vitro* [11].

7.3 Analysis of the size and composition of ribonucleoproteins

7.3.1 Principles of analytical methods

The methods used to determine the size of nucleoproteins are similar to those used for RNA, that is, gel electrophoresis and rate-zonal centrifugation as described in Chapter 3. Modifications to the

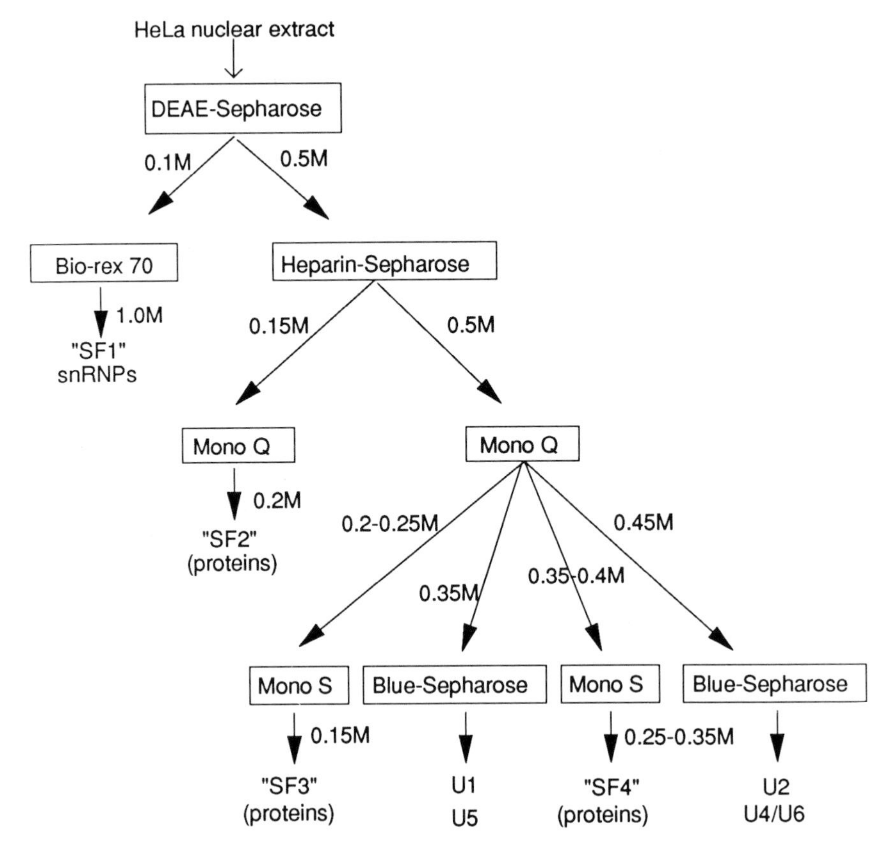

Figure 7.4: Scheme for the chromatographic separation of snRNPs. Molarities shown indicate KCl concentrations necessary for elution.

procedures take into account the different type of particle. The most important criterion is to ensure that whichever method is used the composition is not altered by the analytical procedure.

7.3.2 Use of centrifugation to measure sizes of RNPs

The classical method for determining the size of nucleoproteins is rate-zonal centrifugation on linear sucrose gradients. Molecular sizes are assessed by use of known molecular markers that span the range of sizes of the sample RNPs. The use of sucrose gradients to purify polysomes has been described previously (Section 7.2.2). Good resolution can be achieved using this method which has been developed particularly for polysomes; an example of the type of separation that can be obtained is shown in *Figure 7.2*.

In order to maximize the resolution of gradients, the sample homogenate (Section 7.2.2) should be loaded on to the gradient in a solution which does not contain sucrose as this allows zone sharpening as the nucleoproteins enter the gradient [6]. After centrifugation, the gradient is unloaded into 20–30 fractions and the position of the nucleoproteins determined by UV spectrophotometry at 260 nm. For preparative work, the nucleoproteins can be recovered from the gradient fractions by pelleting. One note of caution is that excessive centrifugal force can disrupt structures such as ribosomes, as a result of the hydrostatic pressures generated towards the bottom of the gradient [5, 6], and this should be kept in mind when deciding upon the centrifugation conditions to be used.

7.3.3 Centrifugal determination of the composition of nucleoproteins

It is frequently easier to measure the relative amounts of RNA and protein in the complex on the basis of their different densities in gradient media than by electrophoretic procedures. This can be achieved by isopycnic centrifugation. The equation used for determining the composition of any complex from its density (D_c) is:

$$D_c = F_r.D_r + F_p.D_p$$

which can be solved for F_r and F_p:

$$F_r = \frac{D_c - F_p D_p}{D_r}$$

and $F_p = \dfrac{D_c - F_r D_r}{D_p}$ assuming $F_r + F_p = 1$

where F_p and F_r are the fractional amounts of protein and RNA, respectively, and D_p and D_r are the densities of protein and RNA in that medium, respectively. Note that the density of RNA can vary with the gradient medium, in CsCl it is 1.9 g/ml, whilst in a non-ionic medium such as Nycodenz or Metrizamide the density is only 1.2 g/ml, depending on the ionic composition of the gradient. The presence of divalent ions in Nycodenz and Metrizamide gradients has a differential effect on ribonucleoproteins making ribosomes in particular very dense. The density of proteins is much less variable at about 1.3 g/ml in most media but this may vary with modification of the protein by the addition of lipid or carbohydrate material. In order to band nucleoproteins in ionic media such as CsCl it is necessary to fix the nucleoproteins with 1–4% (w/v) formaldehyde on ice for 24 h before putting them into the gradient solution (Section 7.2.3).

Nucleoproteins can be banded in non-ionic gradient media such as Metrizamide or Nycodenz without any need for fixation [12]. However, in such media any loosely bound proteins may become detached from the complex [6]. In addition, the presence of Mg^{2+} (necessary for some complex formations) bound to the complexes can modify their buoyant density and in such cases the previous equation is more difficult to apply. Note that once complexes have been fixed it is extremely difficult to analyze them as complexes and to resolve their components further. Therefore there are real advantages in working with unfixed complexes if at all possible.

7.3.4 Use of gel electrophoresis to measure sizes and composition of RNPs

The gel electrophoretic separation of bacterial polyribosomes (polysomes) and ribosomes was made possible by the development of composite gels containing both agarose and polyacrylamide (*Figure 7.5*). The addition of 0.5% (w/v) agarose to low-percentage polyacrylamide gels forms a mechanically stable yet very porous gel. Gels composed of 3.0% (w/v) polyacrylamide and 0.5% (w/v) agarose can be handled easily, and gels with even lower polyacrylamide concentrations can be used if necessary. The high resolution provided by these gels permits the separation of large macromolecules which differ only in conformation or by the presence of a single ribosomal protein [13]. The separation is on the basis of molecular sieving and the extent of separation is determined primarily by the structure (size and shape) of the particle rather than by its charge.

The structure of ribosomes and polysomes depends on the surrounding ionic conditions, particularly the Mg^{2+} concentration. The inert

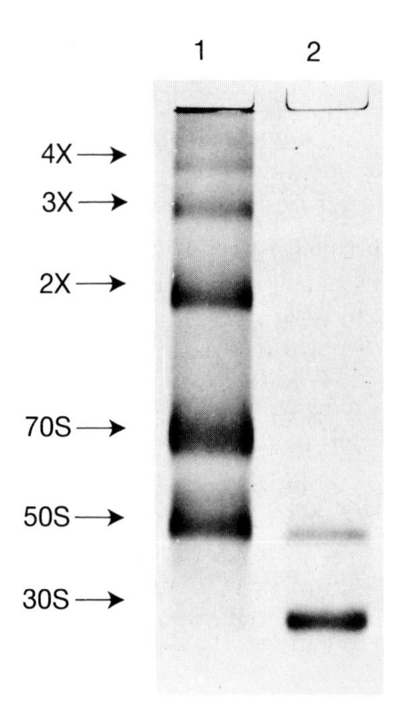

Figure 7.5: Separation of bacterial polysomes and ribosomal subunits by gel electrophoresis. A composite gel of 2.5% polyacrylamide, 0.5% agarose was used to separate (1) a bacterial cell lysate and (2) ribosomal subunits using a running buffer consisting of 25 mM Tris-HCl (pH 7.6), 6 mM KCl, 2 mM $MgCl_2$. Electrophoresis was for 4 h at 4°C at a voltage of 200 V. Reproduced from Dahlberg and Grabowski (1990) in *Gel Electrophoresis of Nucleic Acids: a Practical Approach,* 2nd edn (Rickwood and Hames, eds) with permission from Oxford University Press.

matrix of agarose–polyacrylamide composite gels is particularly well-suited for permitting variations in the ionic conditions so as to allow the study of the ribosome structure. As an example, polysomes from bacteria, which migrate intact in a gel prepared in a buffer containing 1–10 mM magnesium ions, will dissociate and migrate as 30S and 50S ribosomal subunits if the magnesium ion concentration is reduced to 0.2 mM.

The numerous methods available for preparing ribosomes and polysomes have been described in Section 7.2.2, and methods for their electrophoretic analysis are available [14–18].

Protocol 7.6: Electrophoretic analysis of ribosomes and polysomes

1. For optimal resolution of polysomes, the sample should be mixed with an equal volume of warm (50°C) 0.5% (w/v) agarose in buffer and allowed to set in a sample well from which the electrophoresis buffer has been removed. Samples not loaded as a gel segment in a well give a streaked but otherwise identical pattern. About 0.02–0.04 A_{260} units of polysomes and 20–50 µl of 0.25% (w/v) agarose can be mixed and added to each well. The amount of sample loaded is dependent on sample concentration and well size.
2. Gelling of the sample in the well is not necessary for the analysis of ribosomal subunits in gels containing a low concentration (0.2 mM) of magnesium ions as streaking does not occur for these small RNPs. In these cases, the addition of sucrose to a final concentration of 10% (w/v) is sufficient for layering the samples into the wells after the reservoir buffer has been added to the electrophoresis apparatus. Bromophenol blue (0.1% (w/v)) can be added to the samples as a dye marker to visualize the samples during application.
3. Samples should be electrophoresed at a constant voltage of 120–200 V (a voltage gradient of approximately 5–10 V/cm) at 4°C. Ideally the buffer should constantly be recirculated between both electrolyte reservoirs to avoid any changes in pH. Pump systems are often attached to electrophoresis units for this purpose.
4. Immediately after electrophoresis the gels can be stained in Stains-all™ solution (Eastman Organic Chemicals). The gel is placed in a photographic tray containing 10 ml 0.1% (w/v) stock solution of Stains-all, 90 ml 100% formamide and 100 ml distilled water; the tray is covered with aluminum foil and stained overnight. The gels are destained with running tap water while protecting them from direct light due to the

photosensitivity of this stain. Gel strips may be scanned at 570 nm in a densitometer or photographed on a fluorescent viewing box, using a yellow or green filter. The position of the sample bands may be compared to marker band positions to define sizes.

Many analyses of RNP composition now use two-dimensional gel electrophoresis. The first dimension can be run in a 11.4% polyacrylamide tube/rod gel containing 8 M urea at pH 4.5 followed by use of a polyacrylamide slab gel in the second dimension (10% acrylamide, 6 M urea, 0.1% (w/v) SDS at pH 7.1) [16].

Spliceosomes and ribosomes are analogous in that they are both ribonucleoprotein complexes and have enzymatic activity. The similarity also extends to their abilities to be separated by gel electrophoresis. Pikielny and Rosbash [20] were the first to report the use of a composite gel system to study successfully the interaction between snRNPs and pre-mRNA during splicing. They separated yeast spliceosomes by electrophoresis in a 3% acrylamide–0.5% agarose composite gel containing 0.5 × TBE buffer. Other laboratories have also successfully separated mammalian spliceosomes by gel electrophoresis [20,21]. The same problem occurs with electrophoresis of spliceosomes as is seen with mammalian ribosomes; when agarose is present in the gel, aggregation and streaking of samples occurs in composite gels unless EDTA is added to the gel buffer. Consequently, the separation of yeast and mammalian spliceosomes in a composite gel of 3% polyacrylamide/0.5% agarose [20] or 6% polyacrylamide/0.5% agarose [22] has only been reported using 1.25 mM EDTA (0.5 × TBE buffer), and the separation of mammalian spliceosomes in a Tris-glycine buffer without EDTA can only be achieved by electrophoresis in an acrylamide gel without agarose [23]. The ratio of acrylamide to bisacrylamide should be increased considerably in gels for separating spliceosomes to as much as 80:1 in 4% acrylamide gels [23] and 250:1 in 6%/0.5% gels [22] from the normal 20:1.

References

1. **Spedding, G.** (ed.) (1990) *Ribosomes and Protein Synthesis: a Practical Approach.* IRL Press, Oxford.
2. **Higgins, S.J. and Hames, B.D.** (eds) (1993) *RNA Processing: a Practical Approach,* Vol. I. IRL Press, Oxford.
3. **Higgins, S.J. and Hames, B.D.** (eds) (1993) *RNA Processing: a Practical Approach,* Vol. I. IRL Press, Oxford.
4. **Spedding, G.** (1990) in *Ribosomes and Protein Synthesis: a Practical Approach* (G. Spedding, ed.). IRL Press, Oxford. p. 1.
5. **Jelenec, P.C.** (1980) *Anal. Biochem.,* **105**, 369.
6. **Rickwood, D.** (1992) in *Preparative Centrifugation: a Practical Approach.* (D. Rickwood, ed.). IRL Press, Oxford. p. 143.

7. **Jackson, V.** (1978) *Cell,* **15**, 945.
8. **Ip, Y.T., Jackson, V., Meier, J. and Chalkley, R.** (1988) *J. Biol. Chem.,* **263**, 14044.
9. **Dignam, J.D., Lebovitz, R.M. and Roeder, R.G.** (1983) *Nucl. Acids Res.* **11**, 1475.
10. **Kramer, A., Frick, M. and Keller, W.** (1987) *J. Biol. Chem.,* **262**, 17630.
11. **Lamond, A.I. and Sproat, B. S.** (1993) in *RNA Processing: a Practical Approach,* Vol. I (S.J. Higgins and B.D. Hames, eds). IRL Press, Oxford. p. 103.
12. **Ford, T.C. and Rickwood, D.** (1983) in *Iodinated Density Gradient Media: a Practical Approach* (D. Rickwood, ed.). IRL Press, Oxford. p. 23.
13. **Dahlberg, A.E., Dingman, C.W. and Peacock, A.C.** (1969) *J. Mol. Biol.,* **41**, 139.
14. **Peacock, A.C. and Dingman, C.W.** (1968) *Biochemistry,* **7**, 668.
15. **Dahlberg, A.E., Lund, E. and Kjeldgaard, N.O.** (1973) *J. Mol. Biol.* **78**, 627.
16. **Tokomatsu, H., Strycharz, W. and Dahlberg, A.E.** (1981) *J. Mol. Biol.,* **152**, 397.
17. **Dahlberg, A.E., Dahlberg, J.E., Lund, E., Tokimatsu, H., Rabson, A.B., Calvert, P.C., Reynolds, F. and Zahalak, M.** (1978) *Proc. Natl Acad. Sci. USA,* **75**, 3598.
18. **Dahlberg, A.E.** (1974) *J. Biol. Chem.,* **249**, 7673.
19. **Datta, D.B., Changchien, L., Nierras, C.R., Strycharz, W.A. and Craven, G.R.** (1988) *Anal. Biochem.,* **173**, 241.
20. **Pikielny, C.W. and Rosbash, M.** (1986) *Cell,* **45**, 869.
21. **Pikielny, C.W., Rymond, B.C. and Rosbash, M.** (1986) *Nature,* **324**, 341.
22. **Christofori, G., Frendewey, D. and Keller, W.** (1987) *EMBO J.,* **6**, 1747.
23. **Konarska, M.M. and Sharp, P.A.** (1987) *Cell,* **49**, 763.

Appendix A

Suppliers

American Type Culture Collection (ATCC), 12301 Park Lawn Drive, Rockville, MD 20852, USA.
Amersham Corporation, 2636 South Clearbrook Drive, Arlington Heights, IL 60005, USA.
Amersham International plc, Lincoln Place, Green End, Aylesbury, Bucks HP20 2TP, UK.
Amicon Division, WR Grace & Co., 72 Cherryhill Drive, Beverley, MA 01915–1065, USA.
Amicon Ltd, Upper Mill, Stonehouse, Gloucester GL10 2BJ, UK.
Applied Biosystems Inc., 850 Lincoln Center Drive, Foster City, CA 94404, USA.
Applied Biosystems Ltd, Kelvin Close, Birchwood Science Park North, Warrington WA3 7PB, UK.
J.T. Baker Chemicals Inc., 222 Red School Lane, PO Box 492, Phillipsburg, NJ 08865–9944, USA.
J.T. Baker UK Ltd, Wyvols Court, Basingstoke Road, Swallowfield, nr Reading, Berks RG7 1PY, UK.
Barnstead Instrument Co., 2555 Kerper Boulevard, Dubuque, IA 52001, USA.
Beckman Instruments Inc., PO Box 3100, 2500 Harbor Boulevard, Fullerton, CA 92634, USA.
Beckman Instruments UK Ltd, Progress Road, Sands Industrial Estate, High Wycombe, Bucks HP12 4JL, UK.
Becton Dickinson Labware, 2 Bridgewater Lane, Lincoln Park, NJ 07035, USA.
Becton Dickinson Ltd, Between Towns Road, Cowley, Oxford OX4 3LY, UK.
Bellco Glass Co., PO Box 8, 340 Endrido Road, Vineland, NJ 08360, USA. c/o Scientific Laboratories Supplies, Unit 27, Nottingham South and Wilford Industrial Estate, Ruddington Land, Wilford, Nottingham NG11 7EP, UK.
Bethesda Research Laboratories, *see* Gibco BRL.
Bio-Rad Laboratories, Division Headquarters, 3300 Regatta Boulevard, Richmond, CA 94804, USA.
Bio-Rad Laboratories Ltd, Maylands Avenue, Hemel Hempstead, Herts HP2 7TD, UK.
Boehringer Mannheim Corporation, Biochemical Products, PO Box 50414, Indianapolis, IN 46250–0413, USA.

Boehringer Mannheim GmbH Biochemica, PO Box 31 01 20, D-6800 Mannheim, Germany.

Boehringer Mannheim UK (Diagnostics/Biochemicals) Ltd, Bell Lane, Lewes, East Sussex BN7 1LG, UK.

Braun (B. Braun Medical Ltd), 13–14 Farmborough Close, Aylesbury Vale Industrial Park, Bucks HP20 1DQ, UK.

Braun (B. Braun Melsungen AG), PO Box 120, D-3508 Melsungen, Germany.

Brinkman Instrument Co., Cantigue Road, Westbury, NY 11590, USA. c/o Chemlab Instruments Ltd, Hornminster House, 129 Upminster Road, Hornchurch, Essex, UK.

Calbiochem, PO Box 12087, San Diego, CA 92112–4180, USA.

Calbiochem–Novabiochem (UK) Ltd, 3 Heathcoat Building, Highfields Science Park, University Boulevard, Nottingham NG7 2QJ, UK.

Clontech Laboratories Inc., 4030 Fabian Way, Palo Alto, CA 94303–4607, USA.

Corning Medical and Scientific Glass Co., Medfield, MA 02052, USA.

Ciba Corning Diagnostics, Colchester Road, Halstead, Essex CO9 2DX, UK.

Difco Laboratories, PO Box 331058, Detroit, Michigan 48232–7058, USA.

Difco Laboratories Ltd, Central Avenue, East Molesey, Surrey KT8 0SE, UK.

Du Pont Co. (Biotechnology Systems Division), PO Box 80024, Wilmington, DE 19880–0024, USA.

Du Pont UK Ltd, Wedgwood Way, Stevenage, Herts SG1 4QN, UK.

Eastman Kodak Co., PO Box 92822, LRPD–1001 Lee Road, Rochester, NY 14692–7073, USA.
c/o Phase Separations Sales, Deeside Industrial Park, Deeside, Clwyd CH5 2NU, UK.

Falcon, *contact* Becton Dickinson.

Fisher Scientific Co., 50 Fadem Road, Springfield, NJ 07081, USA.

Flow (ICN Biomedical Inc.), 3300 Highland Avenue, Costa Mesa, CA 92626, USA.

Flow (ICN Flow), Eagle House, Peregrine Business Park, Gomm Road, High Wycombe HP13 7DL, UK.

Fotodyne Inc., 16700 West Victor Road, New Berlin, Wisconsin, USA.

Gibco BRL (Life Technologies Inc.), 3175 Staler Road, Grand Island, NY 14072–0068, USA.

Gibco BRL (Life Technologies Ltd), Trident House, Renfrew Road, Paisley PA3 4EF, UK.

Gilson France SA, BP 45, 72 rue Gambetta, 95400 Villiers le Bel, France. c/o Anderman Ltd, 20 Charles Street, Luton, Beds LU2 0EB, UK.

Hoefer Scientific Instruments, PO Box 77387–0387, 654 Minnesota Street, San Francisco, CA 94107, USA.

Hoefer UK Ltd, Newcastle, Staffs ST5 0TW, UK.

HyClone Laboratories, 1725 State Highway 89–91, Logan, UT 84321, USA.

IBI (International Biotechnologies Inc.), PO Box 9558, New Haven, CT 06535, USA.
IBI Ltd, 36 Clifton Road, Cambridge CB1 4ZR, UK.
ICN Biomedicals, *see* Flow (ICN).
ISCO Inc., PO Box 5347, Lincoln, NE 68505, USA.
c/o Jones Chromatography Ltd, Tir-y-Berth Industrial Estate, New Road, Hengoed, Mid Glamorgan CF8 8AU, UK.
Isolab Inc., PO Box 4350, Akron, OH 44398–6003, USA.
c/o Genetic Research Instruments Ltd, Gene House, Dunmow Road, Felstead CM6 3LD, UK.
Kimble Products, 1022 Spruce Street, Vineland, NJ 08360, USA.
Kodak, *see* Eastman Kodak.
Kontes Glass Co., Vineland, NJ 08360, USA.
c/o Burkard Scientific Sales, PO Box 55, Uxbridge, Middx UB8 2RT, UK.
Life Technologies Inc., PO Box 6009, Gaithersburg, MD 20877, USA.
M.A. Bioproducts (Microbiological Associates), Biggs Ford Road, Building 100, Walkersville, MD 21793, USA.
c/o Lab Impex Ltd, Waldergrove Road, Teddington, Middx TW11 8LL, UK.
Macherey Nagel, Postfach 307, Neumann-Neanderstrasse, D-5160 Duren, Germany.
c/o Camlab Ltd, Nuffield Road, Cambridge CB4 1TN, UK.
Millipore Intertech, PO Box 255, Bedford, MA 01730, USA.
Millipore UK Ltd, The Boulevard, Blackmoor Lane, Watford, Herts WD1 8YW, UK.
Nalge Co., PO Box 20365, Rochester, NY 14602–0365, USA,
c/o FSA Laboratory Supplies, Bishop Meadow Road, Loughborough, Leics LE11 0RG, UK.
New England Biolabs (NBL), 32 Tozer Road, Beverley, MA 01915–5510, USA.
New England Nuclear (NEN), Du Pont Co., NEN Research Products, 549 Albany Street, Boston, MA 02118, USA.
Du Pont UK Ltd, Wedgwood Way, Stevenage, Herts SG1 4QN, UK.
Pharmacia Biosystems Ltd (Biotechnology Division), Davy Avenue, Knowlhill, Milton Keynes MK5 8PH, UK.
Pharmacia–LKB Biotechnology Inc., PO Box 1327, 800 Centennial Avenue, Piscataway, NJ 08855–1327, USA.
Pierce, PO Box 117, 3747 North Meridan Road, Rockford, IL 61105, USA.
Pierce Europe BV, PO Box 1512, 3260 BA Oud-Beijerland, The Netherlands.
Progen Industries Ltd, 2806 Ipswich Road, Darra, Qld 4076, Australia.
Promega, 2800 Woods Hollow Road, Madison, WI 53711–5399, USA.
Promega Biotech, 2800 S. Fish Hatchery Road, Adison, WI 53711, USA.
Promega Ltd, Delta House, Enterprise Road, Chilworth Research Centre, Southampton SO1 7NS, UK.

Qiagen Gmbh, Max-Volmer Strasse 4, 4027 Hilden, Germany.
Qiagen Inc., 960 De Soto Avenue, Chatsworth, CA 91311, USA.
Sartorius AG, Postfach 32–43, Weender Landstrasse 94–108, D-3400 Göttingen, Germany.
Sartorius Ltd, Longmead, Blenheim Road, Epsom, Surrey KT9 9QN, UK.
Sartorius North America Inc., 140 Wilbur Place, Bohemia, Long Island, NY 11716, USA.
Schleicher & Schuell, Postfach 4, D-3354 Dassell, Germany.
c/o Anderman & Co. Ltd, 145 London Road, Kingston-upon-Thames, Surrey KT2 6NH, UK.
Searle, *contact* Amersham.
Serva Feinbiochemica GmbH, Postfach 105260, 6900 Heidelberg, Germany.
Sigma Chemical Co. Ltd, Fancy Road, Poole, Dorset BH17 7NH, UK.
Sigma Inc., PO Box 14508, St Louis, MO 63178, USA.
Sorvall, *contact* Du Pont.
Squibb Pharmaceuticals, 1 Squibb Drive, Cranberry, NJ 08512–9579, USA.
Stratagene Inc., 11011 North Torrey Pines Road, La Jolla, CA 92037, USA.
Stratagene Ltd, Unit 140, Cambridge Innovation Centre, Milton Road, Cambridge CB4 4FG, UK.
United States Biochemical (USB) Corporation, Box 22400, Cleveland, OH 44122, USA.
c/o Cambridge Bioscience Ltd, 25 Signet Court, Stourbridge Common Business Centre, Swans Road, Cambridge CB5 8LA, UK.
University of Wisconsin, Genetics Computer Group, University Avenue, Madison, WI 53706, USA.
UV Products Inc., 5100 Walnut Grove, San Gabriel, CA 91778, USA.
UV Products Ltd, Science Park, Milton Road, Cambridge CB4 4BN, UK.
Van Waters & Rogers, PO Box 6016, Cerritos, CA 90702, USA.
Virtis Co. Inc., Route 208, Gardiner, NY 12525, USA.
c/o Damon/IEC (UK) Ltd, Unit 7, Lawrence Way, Brewers Hill Road, Dunstable, Beds LU6 1BD, UK.
VWR Scientific Products, PO Box 7900, San Francisco, CA 94120, USA.
Wako Pure Chemicals, Dosho-Machi, Osaka, Japan.
Waring Commercial, c/o Christison Scientific Equipment Ltd, Albany Road, East Gateshead Industrial Estate, Gateshead NE8 3AT, UK.
Whatman Scientific Ltd, Whatman House, St Leonards Road, Maidstone, Kent ME16 0LS, UK.
Worthington Biochemical Corporation, Halls Mill Road, Freehold, NJ 07728, USA.
c/o Cambridge Bioscience Ltd, 25 Signet Court, Stourbridge Common Business Centre, Swans Road, Cambridge CB5 8LA, UK.

Appendix B

Glossary

Affinity chromatography: fractionation of a specific molecular species from a diverse mixture by use of a molecule showing high affinity to the target bound to a support or bead, e.g. a complementary oligonucleotide bound to cellulose or a magnetic bead.

Anion-exchange chromatography: separation on the basis of binding to a positively charged support matrix and progressive elution by a solution of increasing ionic strength.

Autoradiography: method of detection of radiolabeled components. Autoradiography can be divided into three different techniques: direct autoradiography, fluorography and indirect autoradiography. See individual techniques below for further details.

Biotinylation of oligonucleotides: method of incorporating biotin into oligonucleotides which can be performed using enzymatic or chemical processes. Biotin can then be used to affinity select or mark a specific RNA species.

Blotting of gels: the transfer of nucleic acids (or proteins) to an inert membrane in order to allow detection of specific species.

Cap – mRNA: *see* mRNA – cap.

cDNA: the DNA complement of an RNA strand.

Cell-free translation system: method of synthesizing proteins using extracts isolated from cell types such as wheat germ or rabbit reticulocytes.

Chemiluminescence: light emission produced by enzymatic cleavage of a substrate. The enzyme can be joined to a marker to label specific RNA species.

Denaturing agents: agents which unfold quaternary, tertiary and secondary structures of macromolecules usually by disrupting ionic and hydrogen bonds.

Dideoxy chain termination sequencing: incorporation of terminating nucleotides into growing strands of DNA which can be analyzed to show all possible termination positions and thus the sequence of the template DNA.

Direct autoradiography: labeled components are detected by placing the sample (e.g. dried gel, filter, etc.) in contact with one of the several commercial brands of 'direct' X-ray film (e.g. Amersham Hyperfilm β-max).

Editing of RNA: modification of RNA sequences after synthesis usually by deletion or insertion of nucleotides.

Electrophoresis: the term electrophoresis is applicable to the movement of small ions and charged macromolecules in solution under the influence of an electric field. Many factors affect their rate of migration, these include: the size and shape of the molecule, the charge carried, the applied current and the resistance of the medium.

Electrophoresis buffer: also referred to as electrode, reservoir, running and tank buffer, and electrolytes. Solution required to carry net electrical charge through the gel matrix during electrophoresis.

Equilibration buffer: solution required to equilibrate first-dimensional gels before electrophoresis on the second-dimensional gel matrix. Equilibration buffers are used to add SDS or urea to the gel matrix, and to remove ampholytes and electrolytes present in the first-dimensional gel.

Expression: the production of a gene product from a gene.

Fluorography: detection is achieved by impregnating the gel with a scintillator, such as PPO. The β-particles then interact with the scintillator to cause the emission of light which exposes blue-sensitive X-ray film (eg. Amersham Hyperfilm-MP) placed next to the gel, forming a detectable image.

Gel retardation analysis: when a mixture of protein–nucleic acid complexes and unbound molecules is subjected to gel electrophoresis, the free nucleic acid will rapidly enter the gel as a band, physically removed from the other components. The highly negatively charged RNA will pull bound proteins into the gel, but the RNA–protein complexes are retarded in moving through the gel matrix.

HPLC: a rapid fractionation procedure capable of very high resolution that can be analytical or preparative in scale.

Hybridization: formation of a complex between molecules that have a high affinity for each other, e.g. complementary strands of nucleic acid.

Immunodepletion: removal of a specific member of a molecular complex by immunopurification in order to study the importance of individual members to the function of the complex.

Immunopurification: fractionation of a specific species using a matrix-bound specific antibody.

Indirect autoradiography: the energy of many of the β-particles is such that the emissions pass completely through the X-ray film. However, they can be trapped by placing a calcium tungstate intensifying screen on the other side of the X-ray film. Emissions reaching this screen produce multiple flashes of light which cause the production of a photographic image on the X-ray film, superimposed over the autoradiographic image.

***In vitro* transcription:** synthesis of an RNA molecule using a DNA template.

***In vitro* translation:** synthesis of a protein using an RNA template and a cell-free translation system.

Isoelectric point (pI): the pH at which the net charge on an amphoteric macromolecule is zero.

Loading buffer: *see* Sample loading buffer.

mRNA – cap: eukaryotic mRNA has a dinucleotide cap structure where the two nucleotides bond in a manner that is thought to confer a high stability on the molecule and be involved in ribosome recognition.

Northern blotting: the transfer and immobilization of RNA to a membrane with subsequent probing using DNA probes.

Nuclease S1 analysis: digestion of unprotected areas of a DNA:RNA hybrid to determine RNA termini.

Nucleases: enzymes which degrade nucleic acids. Nucleases can be subdivided into either deoxyribonucleases (DNases) or ribonucleases (RNases). DNases are not generally known for their robust nature, unlike RNases which can be very difficult to inhibit.

Oligo(dT) cellulose: oligonucleotide of thymidylate residues bound to a cellulose molecule. Used for affinity selection of eukaryotic mRNA which have poly(A) tails.

PCR: a powerful technique for exponential amplification of DNA templates.

Polyadenylation: the process of addition of many (50–200) adenylate residues to the 3′ end of eukaryotic mRNA.

Polysomes: complexes of ribosomes with mRNA; the state in which most ribosomes are usually present in the cell.

Proteases: proteolytic enzymes which degrade proteins to peptides and amino acids, either intracellularly or extracellularly.

Relative mobility (R_f): the mobility of the protein or nucleic acid of interest with reference to a marker protein or tracking dye. R_f = distance migrated by protein/distance migrated by dye. The value of R_f is always equal to or less than unity.

Reversed–phase chromatography: fractionation on the basis of the lipophilic nature of nucleic acid bases. Involves initial binding to a non-polar matrix and elution by organic solvents.

Reverse transcriptase: the enzyme responsible for synthesis of a cDNA molecule using an RNA template.

Ribozymes: RNA molecules that possess enzymatic activity, usually associated with proteins.

RNA fingerprinting: two-dimensional separations of partial nuclease digests of RNA molecules.

Running buffer: *see* Electrophoresis buffer.

Sample loading buffer: samples are applied to the gel matrices using sample loading buffers (also referred to as gel loading buffers) containing solubilizing or denaturing agents, inhibitors of proteases and nucleases, dense media such as sucrose or glycerol, and tracking dyes as appropriate.

Size markers: in order to calibrate gels, to estimate the molecular mass of sample proteins, standard proteins of known molecular mass are used as markers. By comparing the R_f of the sample protein to that of the standard proteins, the molecular mass of the sample protein can be determined.

Snurposomes: protein–snRNA complexes involved in splicing.

Solubilizing agents: agents whose effects result in the disruption of non-covalent/hydrophobic interactions.

Spliceosomes: protein–RNA complexes that mediate mRNA splicing.

Splicing of RNA: the process of removal of a region of RNA (usually a non-coding region) and the joining of the two remaining ends of the molecule.

Transfer buffer: solution which acts as a buffered medium for transfer of macromolecules from a gel matrix to an inert support membrane. Allows passage of current and hence facilitates transfer of macromolecules.

Index

LIVERPOOL
JOHN MOORES UNIVERSITY
AVRIL ROBARTS LRC
TITHEBARN STREET
LIVERPOOL L2 2ER
TEL. 0151 231 4022